Application of Molecular Methods and Raman Microscopy/Spectroscopy in Agricultural Sciences and Food Technology

Editors:
Biljana Vucelić Radović,
Dejan Lazić, Miomir Nikšić

2019
Faculty of Agriculture, University of Belgrade

Published by
Ubiquity Press Ltd.
6 Osborn Street, Unit 2N
London E1 6TD

First published 2019

DOI: https://doi.org/10.5334/bbj

The full text of this book has been peer-reviewed to ensure high academic standards. For full review policies, see http://www.ubiquitypress.com/

Suggested citation:
Vucelić Radović, B., Lazić, D. and Nikšić, M.(eds.) 2019. *Application of Molecular Methods and Raman Microscopy/Spectroscopy in Agricultural Sciences and Food Technology.* London: Ubiquity Press. DOI: https://doi.org/10.5334/bbj. License: CC-BY 4.0

Acknowledgements

This work has been supported by European Commission project "Advancing Research in Agricultural and Food Sciences at Faculty of Agriculture, University of Belgrade – AREA", No. 316004 (FP7-REGPOT-0212-2013–I) and OpenAIRE2020 (project governing the FP7 post-grant open access pilot).

A special gratitude we give to our project coordinator Professor Radmila Stikić, whose enthusiasm, efforts and contribution in stimulating disscusions and encouragement, lead us successfully through the course of the AREA project and especially in preparation of this book.

The book has also benefitted from the insightful peer review comments provided by Olivier Piot and Smilja Teodorović.

Contents

Contributors

Ivana Petrović
Faculty of Agriculture. University of Belgrade, Belgrade, Serbia
E-mail: ivana.petrovic@agrif.bg.ac.rs

DraganaBožić
Faculty of Agriculture, University of Belgrade, Belgrade, Serbia
E-mail: dbozic@agrif.bg.ac.rs

MarkolaSaulić
Institute PKB Agroekonomik DOO Beograd, Padinska Skela, Serbia
E-mail: markolasaulic@gmail.com

Sava Vrbničanin
Faculty of Agriculture, University of Belgrade, Belgrade, Serbia
E-mail: sava@agrif.bg.ac.rs

Zorica Ranković-Vasić
Faculty of Agriculture, University of Belgrade, Belgrade, Serbia
E-mail: zoricarv@agrif.bg.ac.rs

Dragan Nikolić
Faculty of Agriculture, University of Belgrade, Belgrade, Serbia
E-mail: nikolicd@agrif.bg.ac.rs

Igor Kljujev
Faculty of Agriculture, University of Belgrade, Belgrade, Serbia
E-mail: ikljujev@agrif.bg.ac.rs

Ivana Stanković
Faculty of Agriculture, University of Belgrade, Belgrade, Serbia
E-mail: ivana.stankovic@agrif.bg.ac.rs

Ana Vučurović
Faculty of Agriculture, University of Belgrade, Belgrade, Serbia
E-mail: ana.vucurovic@yahoo.com

Milan Ivanović
Faculty of Agriculture, University of Belgrade, Belgrade, Serbia
E-mail: milanivanovic@agrif.bg.ac.rs

Nemanja Kuzmanović
Julius Kühn-Institut, Braunschweig,Germany.
E-mail: kuzmanovic1306@gmail.com

Nevena Zlatković
Faculty of Agriculture, University of Belgrade, Belgrade, Serbia
E-mail: nevena blagojevic@agrif.bg.ac.rs

Milica Pavlićević
Faculty of Agriculture, University of Belgrade, Belgrade, Serbia
E-mail: mpavlicevic@agrif.bg.ac.rs

Biljana Vucelić-Radović
Faculty of Agriculture, University of Belgrade, Belgrade, Serbia
E-mail: bvucelic@agrif.bg.ac.rs

Zorka Dulić
Faculty of Agriculture, University of Belgrade, Belgrade, Serbia
E-mail: zorkad@agrif.bg.ac.rs

Božidar Rašković
Faculty of Agriculture, University of Belgrade, Belgrade, Serbia
E-mail: raskovic@agrif.bg.ac.rs

Saša Marić
Faculty of Biology, University of Belgrade, Belgrade, Serbia
E-mail: sasa@bio.bg.ac.rs

Tone-Kari KnutsdatterØstbye
Nofima (Norwegian Institute of Food, Fisheries and Aquaculture Research)
Ås, Norway

Dejan Lazić
Faculty of Agriculture, University of Belgrade, Belgrade, Serbia
E-mail: dlazic@yahoo.com

Steva Lević
Faculty of Agriculture, University of Belgrade, Belgrade, Serbia
E-mail: slevic@agrif.bg.ac.rs

Danka Radić
EDUCONS University, Faculty of Ecological Agriculture, Sremska Kamenica, Serbia
E-mail: danka.radic81@gmail.com

Jovana Vunduk
Faculty of Agriculture, University of Belgrade, Belgrade, Serbia
E-mail: vunduk@agrif.bg.ac.rs

Aleksandar Nedeljković
Faculty of Agriculture, University of Belgrade, Belgrade, Serbia
E-mail: naleksandarn@gmail.com

Ilinka Pećinar
Faculty of Agriculture, University of Belgrade, Belgrade, Serbia
E-mail: ilinka@agrif.bg.ac.rs

Dragana Rančić
Faculty of Agriculture, University of Belgrade, Belgrade, Serbia
E-mail: rancicd@agrif.bg.ac.rs

Miomir Nikšić
Faculty of Agriculture, University of Belgrade, Belgrade, Serbia
E-mail: mniksic@agrif.bg.ac.rs

Preface

This book was created through the funding from the European Commission to the AREA project. AREA aimed to advance research capacity of the scientific groups at the Faculty of Agriculture, University of Belgrade, by strengthening implementation of the existing and introducing innovative technologies that are common to these groups and relevant for agricultural and food sciences.

The protocols and studies presented here are the results of collaboration between the prestigious European laboratories, specialized in state-of-the-art DNA-based technologies or Raman spectroscopy, and the AREA-affiliated groups. The scientists, who were trained at the premises of collaborative institutions, tested the knowledge of acquired techniques by designing and performing experiments, pertinent to their scientific interests and AREA projects.

This book consists of two parts, dedicated to the application of polymerase chain reaction (PCR) and Raman microscopy/spectroscopy in agricultural sciences and food technology. Although conventional PCR and its variant, real-time quantitative PCR (qPCR) have already been applied in different areas of natural and applied sciences, the use of Raman microscopy/spectroscopy has slowly propagated into agricultural sciences and food technology.

The four sections of the first part of the book present molecular methods for a wide scope of applications in agricultural sciences, and food technology, giving introductory details and, emphasizing mostly the role of qPCR techniques in *plant sciences, microorganisms, food biochemistry and technology, and fishery.*

Each chapter is written by scientists with hands-on experience in these fields and covers specific protocol and step-by-step instructions, including but not limited to: (i) two step RT-qPCR analysis of gene expression in plant organs, (ii) one step RT–PCR detection of phytopathogenic viruses, (iii) multiplex qPCR assay for simultaneous detection of quarantine plant pathogenic bacteria, and (iv) application of qPCR in identification of lineages, markers and loci that are responsible for economically important traits in aquaculture and fishery.

In the second part, the book presents techniques of Raman microscopy/spectroscopy, discusses available software for data manipulation and classification of Raman spectra, and includes a short instruction how to operate one of Raman microscope. A separate chapter is dedicated to a comprehensive list of working conditions that laboratory for Raman microspectroscopy should meet before setting up the sample analysis. Raman microspectroscopy is a safe, sensitive and easy to use analytical technique. Utilizing near-infrared lasers reduces the risk of damaging biological samples that makes this method suitable for *in vivo* study on structure, chemical composition and properties of cells and tissues. In three sections on specific Raman microscopy protocols the focus is on: (i) *microorganisms* – characterization of microorganisms, (ii) *food research* – analysis of dairy products and mushroom extracts, and (iii) *functional crop anatomy* - carotenoids detection in fruit material.

PART I

PCR in agricultural sciences and food technology

Two-step RT-qPCR analysis of expression of 7 drought-related genes in tomato (*Lycopersicon esculentum* Mill.)

Ivana Petrović

Abstract

The identification and characterization of genes induced under drought stress is a common approach to elucidate the molecular mechanisms of drought stress tolerance in plants.Examination of gene expression using quantitative PCR (qPCR) in combination with Reverse Transcription (RT) in plant responses to drought stress can provide valuable information for stress-tolerance improvement. The purpose of this manuscript is to describe procedure for two step RT-qPCR analysis of gene expression in tomato leaves, under controled conditions and under drought stress. Described protocol can be adjusted and used for gene expression analysis of different plant species.

1 Introduction

Climate change is one of the most serious problems facing the agriculture today. In a many countries, drought in conjunction with high temperature becomes a significant risk for sustainable agricultural production. In general, drought stress limits productivity of major crops by inducing different morphological,

How to cite this book chapter:
Petrović, I. 2019. Two-step RT-qPCR analysis of expression of 7 drought-related genes in tomato (*Lycopersicon esculentum* Mill.). In: Vucelić Radović, B., Lazić, D. and Nikšić, M. (eds.) *Application of Molecular Methods and Raman Microscopy/Spectroscopy in Agricultural Sciences and Food Technology*, Pp. 3–14. London: Ubiquity Press. DOI: https://doi.org/10.5334/bbj.a.

physiological and molecular changes in plants (Ashraf et al. 2013). At the molecular level, drought stress induces expression of water-deficit-related genes. The products of those genes allow plants to protect cellular function and to adjust plant metabolism.

Tomato (*Lycopersicon esculentum* Mill.) is one of the most widely grown vegetables in the world. Tomato fruits are of special importance both as a fresh vegetable and as a component of food processing industry. However, most of the commercial tomato cultivars are drought sensitive at all stages of the development, with the seed germination and seedling growth being the most sensitive stages (Foulard et al. 2004). Similarly to many other vegetables, tomato has high water requirements (CA. 400–600 mm ha-1) and water supply is essential for successful production (Hanson & May 2004).

Real-time PCR is a technique that measures quantity of target sequence in real time and that is commonly used toquantify DNA or RNA in a sample. Using sequence-specificprimers, the number of copies of a particular DNA or RNA sequence can be determined. By measuring the amountof amplified product at each stage during the PCR cycle, quantification is possible.SYBR Green-based detection is the least expensive and easiest method available for real-time PCR. SYBR Green specifically binds double-stranded DNA by intercalating between base pairs, and fluoresces only when bound to DNA. Detection of the fluorescent signal occurs during the PCR cycle at the end of either the annealing or the extension stepwhen the greatest amount of double-stranded DNA product is present.

Expression of drought- related genes can reveil the role of their products in drought resistance mechanisms. Those informations can be helpful in the breeding efforts to produce tomato cultivars with the increased/sustained fruit quantity and quality in drought conditions.

2 Materials, Methods and Notes

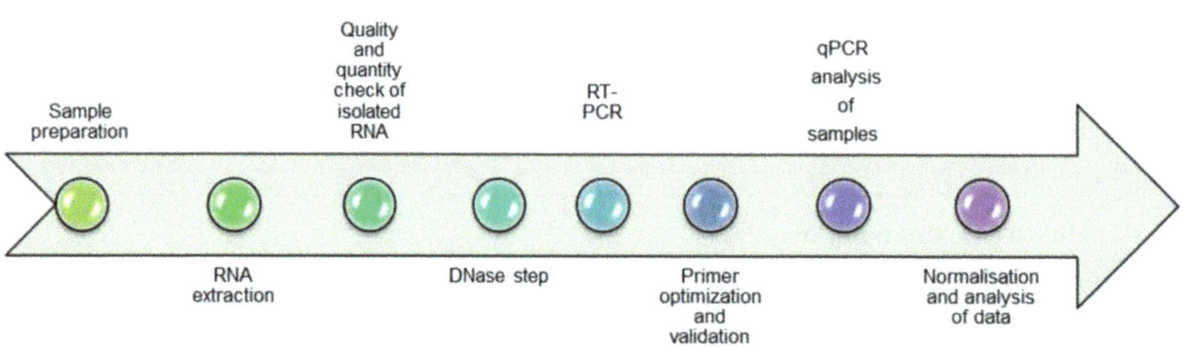

Figure 1: Phases of two-step RT-qPCR.

2.1 Sample preparation – tomato leaves

Note:

- Only young and fully developed leaves should be collected. Old and damaged leaves are not a good material for qPCR analysis of drought-related genes.
- To avoid RNA degradation by RNase, collected samples should not melt at any moment after freezing in liquid nitrogen.
- To avoid cross-contamination, it is necessary to use clean tools for collecting of each leaf and to clean the grinder well after every sample with some DNA/RNA cleaning reagent.

2.1.1 Collect tomato leaves and put them into sterile, unused bags made from liquid-nitrogen proof material. Bags should be placed immediately into liquid nitrogen.
2.1.2. Grind collected leaves in grinder with liquid nitrogen.
2.1.3. Transfer around 150 mg of leaf powder into clean 2 ml tube.
2.1.4. Store tubes at −80°C until analysis.

2.2 RNA extraction

Note:

- Method which includes using of TRIzol REAGENT is one of the most effective methods of RNA isolation. The procedure with TRIzol REAGENT can be completed within 1 hour and the recovery of undegraded mRNAs is 30–150% greater than/ when compared to other methods of RNA isolation. For the extraction from tomato leaves, this method is efficient and RNA has good quality. In this study, TRIzol REAGENT-Thermo Fisher Scientific was used.

The extraction of RNA from tomato leaves is done by following steps:

a) HOMOGENIZATION
   2.2.1. Homogenize tissue samples in TRI Reagent (1 ml/100 mg tissue*). Mix well with vortex.
   2.2.2. Store the homogenate for 5 minutes at room temperature.
   *The sample volume should not exceed 10% of the volume of TRIzol because an insufficient volume can result in DNA contamination of isolated RNA.

b) SEPARATION
   2.2.3. Add 200μl of chloroform per 1 ml of TRI Reagent, cover the samples tightly and shake vigorously for 15 seconds with vortex.

2.2.4. Store the resulting mixture at room temperature for 2–15 minutes.
2.2.5. Centrifuge at maximum speed for 15 minutes at 4 C.
2.2.6. Transfer the 500 μl of the aqueous phase to a new tube.

c) RNA PRECIPITATION
2.2.7. Add 500 μl of isopropanol and mix quickly by inversion.
2.2.8. Store samples at room temperature for 5–10 minutes and centrifuge at max.speed for 10 minutes at 4°C.

d) RNA WASH
2.2.9. Remove the supernatant and wash the RNA pellet (by vortexing) with 1ml 75% ethanol.
2.2.10. Subsequent centrifugation at 10000rpm for 5 minutes at 4°C.

e) RNA SOLUBILIZATION
2.2.11. Remove the ethanol wash and briefly air-dry the RNA pellet for 5–10 min. It is important not to completely dry the RNA pellet because drying will decrease its solubility.
2.2.12. Dissolve RNA in water RNase-free (50μl) by passing the solution a few times through a pipette tip, vortex if necessary.
2.2.13. Store at –20° C for short periods, otherwise store at –80° C.

2.3 Quality and quantity check of isolated RNA

Validation of quality and amount of isolated RNA is required. Quality check can be done by agarose gel electrophoresis. In this study, RNA quality control was done on 1%agarose gel. Into precast gelsmixture of 2μl RNA, 3μl of RNase-free H_2O and 1μl of loading buffer was loaded. General information about RNA integrity can be obtained by observing the staining intensity of the major ribosomal RNA (rRNA) bands and any degradation products*. In this work, total RNA formed clear 28S and 18S rRNA bands (ratio 2:1), which is a good indication that the RNA had good quality.

Quantification of RNAs was done by NanoDrop spectrophotometer and samples were diluted, until concentration of 200 ng of RNA/1 μl of sample was obtained. For extracted RNA, the ration of 260/280 close to 2 indicates the high-quality material, suitable for further analyses.

* Partially degraded RNA will have a smeared appearance, will lack the sharp rRNA bands, or will not exhibit the 2:1 ratio of high quality RNA. Completely ensure the gel was run properly. Degraded RNA will appear as a very low molecular weight smear. Use of RNA size markers on the gel will allow the size of any bands or smears to be determined and will also serve as a good control to

2.4 DNase step

Note:

- Important controle in RT-qPCRanalysis is DNase step, in which the isolated RNA is treated with DNase enzyme. This step ensures that analyzed samples of RNA are clean from genomic DNA contamination that can affect results: The false-positive RT-PCR product could come from the presence of genomic DNA instead of RNA. DNase used in this work was part of the RNase-Free DNase Qiagen kit (ref: 79254).

Before performing DNase step, it is required to do efficacy test of DNase buffer and DNase enzyme. Buffer test and DNase efficacy test are performed with 2–3 fold concentrated samples of RNA, compared to concetration used for RT-qPCR reaction.

Three test tubes should be made:

Tube 0 = 18 μL H20 RNase free + 2 μL RNA
Tube 1 = 16 μL H20 RNase free + 2 μL RNA + 2μL DNase buffer
Tube 2 = 15.8 μL H20 RNase free + 2μL RNA + 2μL DNase buffer + 0.2 μLDNase

2.4.1 DNase buffer test

2.4.1.1. Incubate tubes 0, 1 and 2 during 30 min at 37°C + 5 min at 65°C. The purpose of incubation (at 65°C) is inactivation of DNase, present only in tube 2.
2.4.1.2. Mixture from tubes 0 and 1 should be run on agarose gel, in order to check that DNase buffer did not degrade RNAs.
2.4.1.3. Tubes should be kept at –80°C for DNase test.

Preparation of Tris-HCl (1M pH 8,00)	DNase solution
605,7 mg of Tris	2 ml of 1M Tris-HCl pH=8,00
235μL of 37 % HCL	0,4 ml $MgCl_2$
Adjustement of pH=8,00	0,4 ml DTT (0,1 M) – from DNase kit)
5mL of H_2O	5mL of H_2O
Preparation of $MgCl_2$ 0,5M	↓
508 mg of $MgCl_2$	Filter DNase buffer by 0.22 μM filter
5mL of H_2O	Store at –20°C

Table 1: DNase buffer (5 ml) preparation protocol.

2.4.2 DNase test

Note:

- This test is in fact a real time PCR with a housekeeping gene and SYBR Green as fluorescent probe. The aim is to check if there is still genomic DNA in the purified RNA sample after the DNase step treatment.
- DNase test is done in presence of positive (tomato RNA) and negative (H_2O) control.
- For DNase test, it is recommended to use the products from DNase buffer test (from 2.4.1.) – content from tube 1 (sample without DNase enzyme) and tube 2 (sample with DNase enzyme).

95°C	10 min	1 cycle
95°C	30 sec	40 cycles
55°C	1 min	
72°C	30 sec	

Table 2: Real-time PCR conditions for DNase test.

Results should be checked. There should be no DNA in samples and no PCR products in qPCR reaction.

2.4.3 DNasestep

Note:

- Before DNase step on all samples, it is important to dilute RNA until 2 μg/μl concentrations is obtained. The easiest way is to dilute samples in wells of the plate, so the next step is easier. In this study after dilution each well contained 17.8 μL of diluted RNA.

2.4.3.1. In each well add 2 μL of DNase buffer and 0.2 μL of DNase
2.4.3.2. Incubate 30 minutes at 37°C.
2.4.3.3. Incubate plate for 5 minutes at 65°C in order to inactivate DNase.
2.4.3.4. Store plate at –80°C.

2.5 Two-step RT-qPCR

There are two approaches to RT-qPCR. First one is one-step RT-qPCR that combines the RT reaction and PCR in one plate. Second one is two-step RT-qPCR where the RT reaction is performed separately from the qPCR. In this

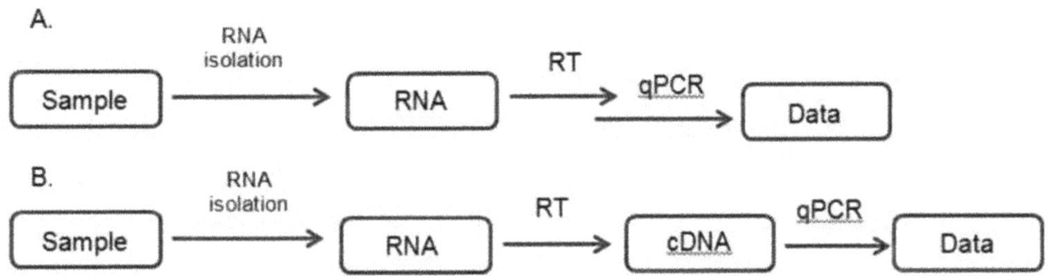

Figure 2: A. One step RT-qPCR B. Two-step RT-qPCR.

study, we used two-step approach because it provides bigger control of processes and higher level of flexibility. This approach also simplifies any required troubleshooting.

2.5.1 RT TEST

Note:

- The aim of this test is to check the efficacy of the buffer and of the DNase during RT-PCR before to make this step on all the samples. For this test, 2–3 samples can be used or a pool of RNA samples. If we have different conditions, it's better to have one pool by condition (in this case, control and drought stress).

		Without superscript	**Without superscript**	**With superscript**	**With superscript**
	H_2O test	**Condition 1**	**Condition 2**	**Condition 1**	**Condition 2**
Oligo (dT)21	1 µl	1 µl		1 µl	
RNA	/	10 µl		10 µl	
dNTP Mix	2,5 µl	2,5 µl		2,5 µl	
H_2O	10 µl	/		/	
Incubation	5 min at 65°C + 5 min on ice				
Buffer (kit)	4 µl	4 µl		4 µl	
DTT (kit)	1 µl	1 µl		1 µl	
Superscript III	0,75 µl	/		0,75 µl	
Incubation	60 min at 42°C + 5 min at 70°C				

Table 3: RT test.

This test is done by RT PCR. After last incubation, results should be checked on agarose gel. On gel should be checked negative controls (H_2O and RT without superscript), and RT product with superscript. Negative controls do not contain DNA, so there should not be present DNA traceson gel. DNA ladders are used in gel electrophoresis to determine the size and quantity of testing DNA fragment. DNA leader can be also used as positive control, to confirm the formation of good smear – one clear band of DNA. If two bands appear, it could indicate that some of the products are single stranded. Presence of big smear indices that DNA is degraded.

2.5.2 RT

If initial RT test (2.5.1.) is successful, the RT procedure should be done for all samples. During this procedure the cDNA of each sample is synthesized. Once cDNA is made, 2 µl of every sample should be mixed into a pool (or multiple pools for multiple conditions) that is going to be used for primer validation.

The rest of cDNA should be stored in plate at –80°C.

	1 sample	50 samples	98 samples	
RNA	10µL	10 µL	10 µL	by well
oligo(dT)21	1 µL	50µL	98 µL	3,5µL by well
dNTP Mix	2,5 µL	125 µL	245 µL	
Incubation	5 min at 65°C + 5 min on ice			
Buffer (kit)	4 µL	200 µL	392 µL	5,75 µL by well
DTT (kit)	1 µL	50 µL	98 µL	
Superscript III	0,75 µL	37,5 µL	73,5 µL	
Incubation	60 min at 42°C + 5 min at 70°C			

Table 4: RT PCR.

2.6 Primer optimization and validation

Primer optimization and validation are essential, even when using primers that have been predesigned and commercially obtained. Optimization is required to ensure that the primer is as sensitive as it is required and that it is specific to the gene of interest.

Primer validation should be carried out on a *pool* of all available *cDNAs* (pool of cDNA made from all analyzed samples). In this study, one *pool of* cDNAs was made from samples exposed to drought stress and second pool is made from control samples. Both pools are diluted with ultra-pure water (10µl of cDNA pool and 90 Ml of ultra-pure water). Dilutions are kept at –20°C. Primers also should be diluted to obtain different concentrations (10^{-3}–10^{-12}). Important data gotten from this step is also primer efficiency.

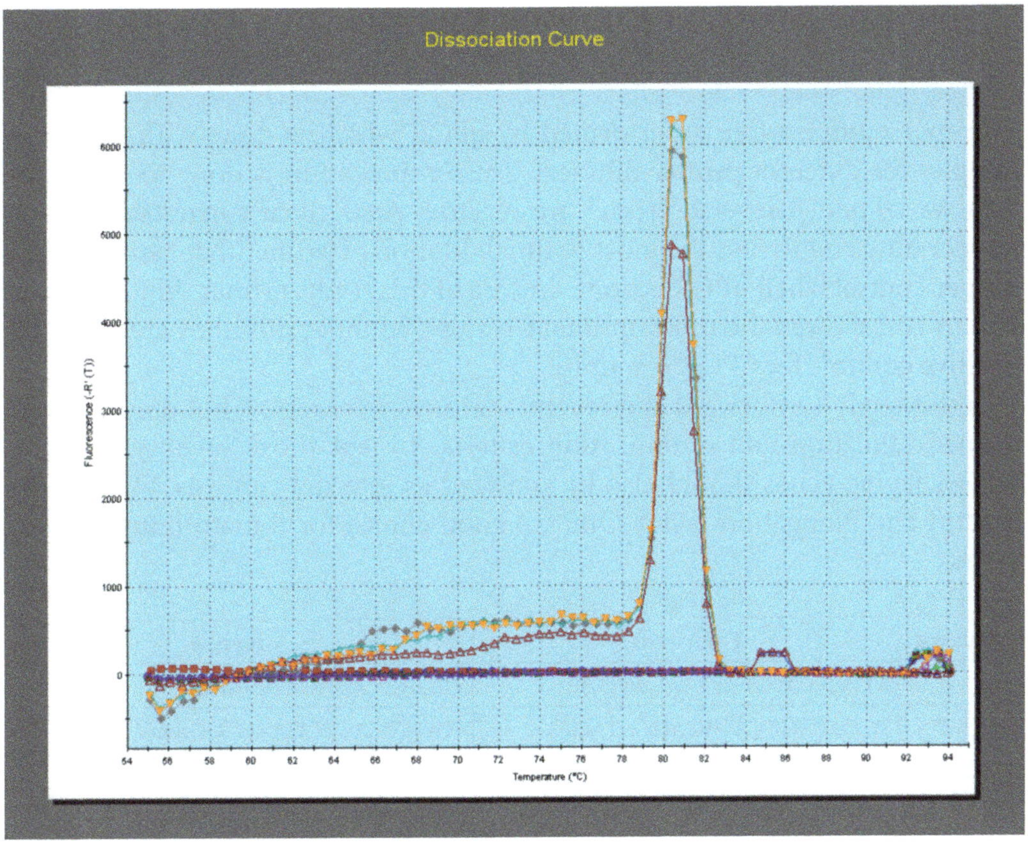

Figure 3: Dissociation peaks of primer with high specificity.

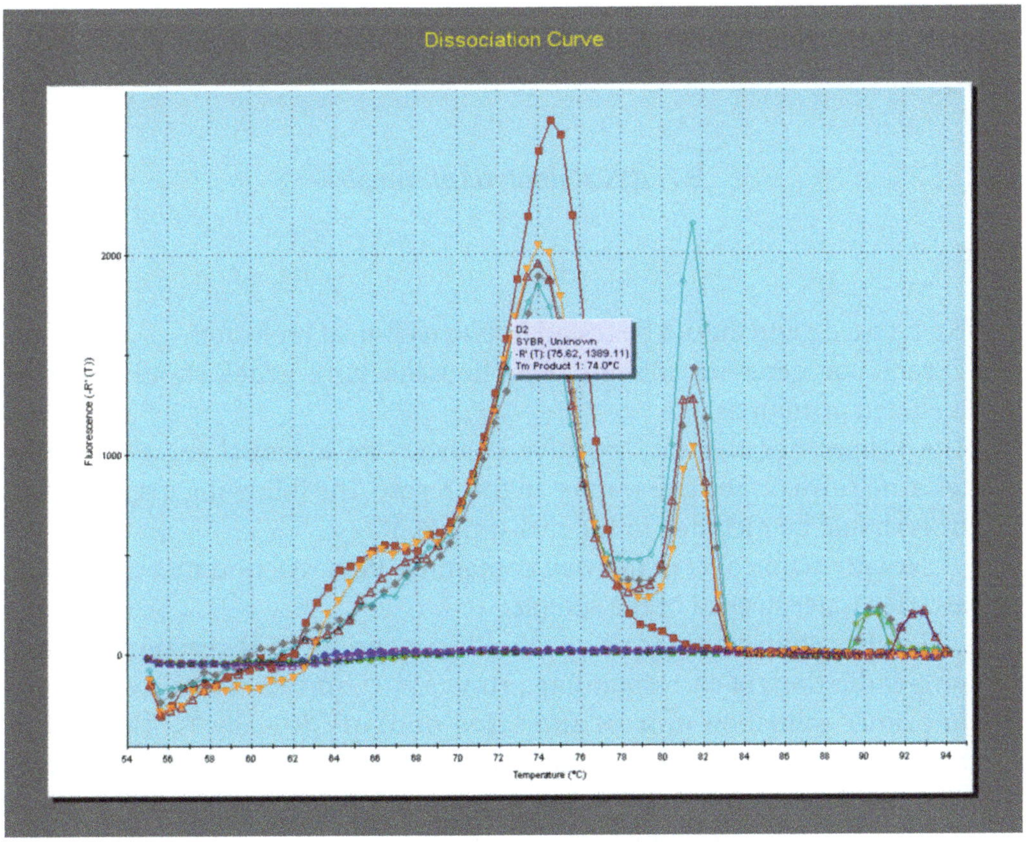

Figure 4: Dissociation peaks of primer with low specificity.

Primer optimization is performed by qPCR which is done with a pool of samples for different primer dilution. This optimization is done to check the F-forwared and R-reverse primer are reacting properly at suggested reaction temperature and to find the most optimal dilution of primer that can be used a proper control when qPCR is done. In case of this study, primer dilutions from 10^{-7} to 10^{-8} showed the most optimal Ctvalues, so those dilutions are saved for positive controls for qPCR reactions.

In this study 12 (forward and reverse) primers were tested, but only 7 passed primer validation and optimization criteria. Except those seven genes, two housekeeping genes should also be analyzed as internal controls. For tomato, β-actin and Elongation factor One are good choice for tomato housekeeping genes.

Primers	F- forward	R-reverse
ZEP1-1	ATCAACTGTGGGAACACCTG	ACGACCAGACATCTGCAATC
ZEP1-2	TGCATGGCCATAGAGGATAG	TGGATGACTCCAACTCGAAG
PPC2	TCAAACTCCACAGTGCGATG	CCGCAATTGGAAACGATG
SlAPXcyto	CCTTTGTGATCCTGCTTTCC	CAGCTCTTCCAATCAGCATC
NCED1	AGGCAACAGTGAAACTTCCATCAAG	TCCATTAAAGAGGATATTACCGGGGAC
SlAPXcp	TTGATCCACCTGAGGGTTTC	TCCCAAGCCTTCGTATTCTG
abi1	GGCAGCAAGGACAACATAAC	TGAGGCCAATTGTGTTGAAG

Table 5: Primers for quantitative real-time PCR (optimized and validated).

2.7 qPCR analysis of samples

Note:

- Each tested gene should be tested in two technical replicates.
- Except our genes of interest, two housekeeping genes should also be included in analysis.
- The proper negative and positive controls are essential for eliminating false-negative or positive results. In this regard, the following negative controls should be included in the real-time PCR test:

  Negative control is in the well containing PCR reaction mix and nuclease-free water instead of the sample.

  Positive controls are in the two wells containing PCR reaction mix and proper dilutions of corresponding primers (10^{-7} and 10^{-8}) that are obtained in primer validation process and saved until qPCR analysis. Those positive controls are needed to validate accuracy of PCR reaction: it is important that values from our primer validation process are similar to those obtained in qPCR reaction with our samples.

– During sample preparation and qPCR analysis, it is important to avoid contamination. If contamination occurs, it is essential to determine the source of contamination. More information about contamination detecting and solving the problem can be found at this link http://www.gene-quantification.com/mifflin-optimisation-report.pdf.

2.7.1. Dilute all samples 1/15 (5 μl of cDNA and 70 μl of ultra-pure water) in the plate, in order in which all samples will be distributed during all analysis
2.7.2. Distribute 2 μl of diluted cDNA into multiple plates. Those plates are "ready to use" and they can be stored at –20° C for short periods.
2.7.3. Distribute 18 μl of Master Mix into "ready to use" plate
2.7.4. Run qPCR and save the results.

To avoid potential contamination, it is desirable to separate samples from controls on qPCR plate (controls should be on the other part of the plate).

Number of wells	1	6
H20	6.2	37.2
Briliant II Sybr Green Master Mix – Agilent Technologies Stratagene	10	60
Rox 1/500	0.3	1.8
primer	1.5	9

Table 6: Reagents mixture for real-time PCR.

95°C	10 min	1 cycle
95°C	30 sec	40 cycles
55°C	40 sec	
72°C	30 sec	
Dissociation curve		

Table 7: Real-time PCR conditions.

After qPCR, amplification plot and dissociation peak should be checked. For each gene, only one dissociation peak should be visible. It means that primer has good specificity. Ct values should be between 15 and 25, which mean that good level of expression is present. After qPCR analysis, it is necessary to do data normalization before statistical analysis.

Data normalization in real-time RT-PCR is one of the major steps in qPCR analysis. Data normalization can be carried out against an endogenous unregulated reference gene transcript or against total cellular DNA or RNA content. In this study, normalization is done by using two internal controls, which are basically two reference housekeeping genes. Transcripts of such genes, which are expressed at relatively high levels in all cells, make ideal positive controls for determining whether or not genes of interest are expressed in given types of samples under given conditions.

It is recommended to use between two and five validated stably expressed reference genes for normalization. It is important to use genes which are validated and which for sure have stable expression. Stability of reference genes can be determined by calculating their M value (M) or their coefficient of variation on the normalized relative quantities (CV). These values can then be compared against empirically determined thresholds for acceptable stability.

Acknowledgements

This work was funded by by EU Commission project AREA, no. 316004. I would like to thank to researchers from INRA (Avognon, France), especially Nadia Bertin and MatildeCausse for arranging our visit and Justine Gricourt for laboratory support.

References

Ashraf, M. & Harris, P.J.C. (2013). Photosynthesis under stressful environments; An overview. *Photosyntetica 51*, 163–190, DOI: https://doi.org/10.1007/s11099-013-0021-6

Foolad, M.R., Zhang, L.P. & Subbiah, P. (2003). Genetics of drought tolerance during seed germination in tomato: inheritance and QTL mapping. *Genome 46*, 536–545. DOI: https://doi.org/10.1139/g03-035

Hanson, B. & May, M. (2004). Effect of subsurface drip irrigation on processing tomato yield, water table depth, soil salinity, and profitability. *Agricultural Water Management. 68*, 1–17. DOI: https://doi.org/10.1016/j.agwat.2004.03.003

Molecular BioProducts (1997). Control of Contamination Associated with PCR and Other Amplification Reactions by Theodore E. Mifflin, Ph.D., DABCC, Retrieved from http://www.gene-quantification.com/mifflin-optimisation-report.pdf

Application of molecular methods in weed science

Dragana Božić, Markola Saulić, Sava Vrbničanin

Abstract

Molecular methods are useful tools for weed science, especially in the area of weed resistance to herbicides and gene flow from herbicide tolerant crops to their wild relatives. Also, genetic variability plays an important role in weed susceptibility to herbicides and affect on strategies of control. For all of these studies, DNA, as a starting material, could be extracted by various methods; though, the easiest and the most suitable is extraction by using commercially available kits. The most important part of molecular analysis is selection and design of adequate primers for successful DNA amplification. Usually, primer selection and designing are based on DNA sequences stored in GenBank. Analysis following selected DNA fragments will depend on type of research. For weed resistance or gene flow studies, amplified fragments are sequenced and obtained information compared with the GenBank sequence database, with the aim to check for mutation(s) presence. For genetic diversity of weed species analysis of amplified DNA fragments include Capillary Electrophoresis.

How to cite this book chapter:

Božić, D., Saulić, M. and Vrbničanin, S. 2019. Application of molecular methods in weed science. In: Vucelić Radović, B., Lazić, D. and Nikšić, M. (eds.) *Application of Molecular Methods and Raman Microscopy/Spectroscopy in Agricultural Sciences and Food Technology*, Pp. 15–22. London: Ubiquity Press. DOI: https://doi.org/10.5334/bbj.b.

1 Introduction

Molecular methods can be useful for different weed science research topics including molecular determination of weed species which is difficult for determination based on non- molecular methods, population variability, weed resistance to herbicides and gene flow between herbicide-tolerant crops and their wild relatives.

Over the last period weed resistance to herbicides has become an increasing problem (Moss et al. 2007; Michitte et al. 2007). In most cases, evolved weed resistance is due to mutation/mutations within gene encoding enzymes which represent herbicide target site. Therefore, it is possible to use a variety of molecular-based assays that are much faster and less labor intensive than traditional methods (e.g. whole-plant bioassay). Detecting weed resistance to herbicides using DNA based techniques is a very important mission, especially with increasing use of newly-bred herbicide- tolerant crops. There are potential risks associated with growing these crops such as gene flow from herbicide-tolerant crops to non-tolerant crops, or to wild relatives, or volunteer crops. This leads to incidences of resistant species (weeds) (Martinez-Ghersa et al. 1997). Molecular markers have proven valuable in determining the frequency of crop-weed hybridization.

Molecular-based approaches have been used in a variety of ways to explore the genetic diversity of weeds. Studies of genetic diversity can be used for determination of species center of origin (Goolsby et al. 2006, Madeira et al. 2007). Such knowledge can be used to direct searches for potential biological control agents (Paterson et al. 2009). Other genetic diversity studies have been conducted with vegetative propagated perennial weed species, with a goal to determine the relative role of sexual versus vegetative reproduction to the success of the weeds (Slotta et al. 2006). Also, studies of genetic diversity in weed populations can be extremely important because they provide essential background for their different susceptibility to herbicides.

2 Materials, Methods and Notes

2.1 Assessment of gene flow from herbicide tolerant sunflower to weedy sunflower by end-point PCR

End-point PCR is suitable for detection mutations responsible for weed resistance to herbicides and confirmation gene flow from tolerant crops to weedy relatives. Namely, the alteration of ALS (acetolactate synthase) gene by one of many possible point mutations is main mechanism of weed resistance to ALS-inhibiting herbicides, which represent group of herbicides to which weeds usually developed resistance. There are eight possible point mutations detected until now in different weed species. As position of potential mutation is known, their detection is based on amplification of the appropriate DNA fragment, using specific primers. After amplification PCR products have to be sequencing and check presence of mutation in obtained sequences.

2.1.1 Plant material

Seed material for weed resistance to herbicides research should be is collect in the fields for which there are indications about resistance development. Also, it is necessary to collect seeds of the same species from the areas where there is no herbicide application history. For gene flow studies, it is possible to establish field experiment which includes different variants of crop-wild relative distance or to collect seeds from wild relatives of crop (in our case weedy sunflower seeds) from the tolerant crop growing area. Young plants produced from collected seeds are sampled and some leaf samples (taken from a single plant) were used for DNA extraction immediately after sampling, while some of them stored in a freezer (–20°C) until analysis. Before analysis samples were lyophilized after storage at the –80°C during 24h.

2.1.2 DNA extraction

DNA extraction was done using the QiagenDneasy® Plant Mini Kit following the manufacturer's protocol (https://www.qiagen.com/dz/shop/sample-technologies/dna/dna-preparation/dneasy-plant-mini-kit/). The quality and concentration of extracted DNA were determined spectrophotometrically using a Nanodrop® 1000. DNA extracts were stored at –20°C when not in use.

Note: The samples were grinded to a fine powder either using a mortar and pestle or TissueLyser. The lyophilized samples were grinded successfully with both methods, but TissueLyser was not a good choice for fresh samples. Instead of getting a fine powder, the TissuLyser was turning fresh material into squashy product.

2.1.3 Primers selection

DNA sequences from several sources were used to design oligonucleotide primers for amplifying ALS gene fragments (White et al., 2003, Kolkman et al., 2004). Two primers were designed using the software Primer 3. The primers Hel ForA (CAATGGAGATCCACCAAGCT) and Hel RevA (AACGCAAGCAACAAATCACT) used for amplification approximately 700bp fragments.

Note: In the literature, there are plenty of primers, which can be used to detect mutations responsible for resistance/tolerance of different sunflower forms to herbicides. Based on their analysis and comparison, and analysis of sequences of DNA fragments from sunflower stored at the GenBank new primers were designed.

2.1.4 Amplification of specific region of the ALS gene

Final PCR reaction condition were: 19 µl of mastermix (10 units Biomix, 7 units DEPC water, 1 unit forward primer and 1 unit reverse primer) and 1µl of DNA sample. Cycling conditions were: 2 min incubation at 94°C; 35 cycles of 30 sec denaturation at 94°C, 20 sec annealing at 53°C and 45 sec extension at 72°C; and 5 min final extension at 72°C. PCR products were electrophoresed on 2% low-melt agarose gel containing ethidium bromide.

Note: Amplification of DNA fragments was successful while using Biomix, purchased from one manufacturer. Switching to other manufacturer of Biomix failed to generate any PCR products. Initially, we didn't realize what was causing the problem and spent significant time and materials, checking other components of PCR reaction and optimizing the assay itself, but without success. Finally, when we changed Biomix again, and chose the one that had been initially used, the amplification became successful again.

2.1.5 Sequencing

PCR products purification was done before sequencing using the Spin Column PCR Purification Kit following the manufacturer's protocol (http://www.nbsbio.co.uk/downloads/DNA_Cleanup_Handbook.pdf). Purified products were sent together with the corresponding primer (Hel ForA) to Sorce Bioscience (Osford, UK) for sequencing. Analysis of obtained sequences were done based

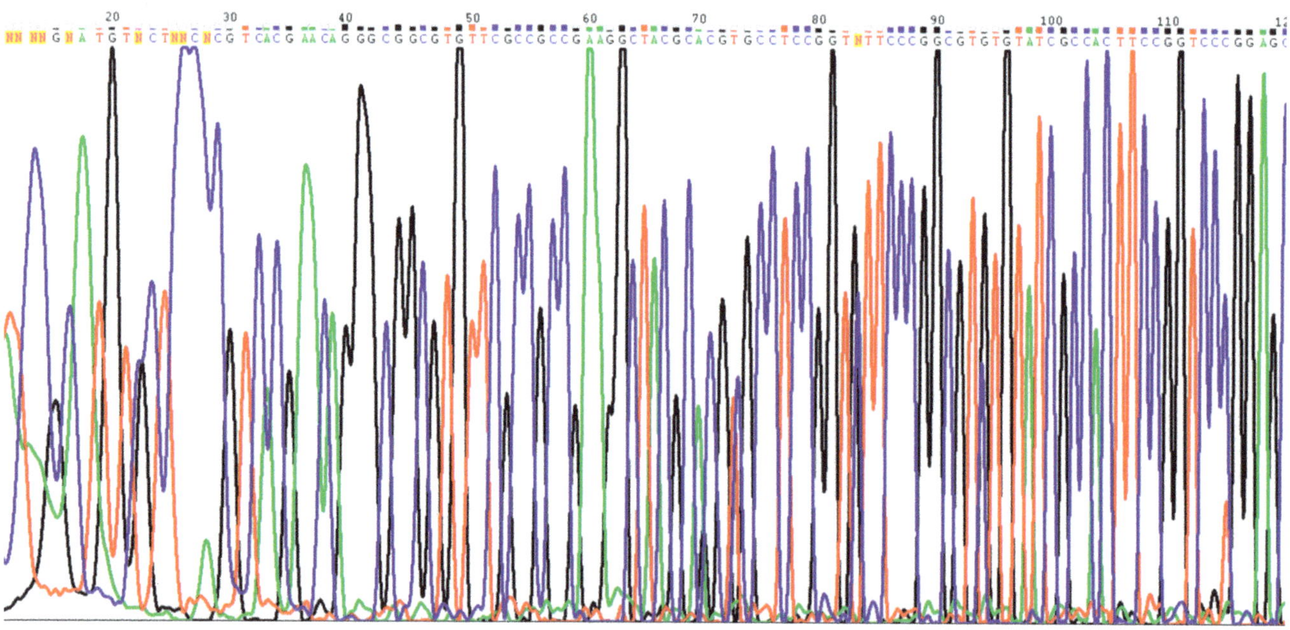

Figure 1: Chromatogram from repeated sequencing of the region of the ALS gene in weedy sunflower DNA.

on comparison with sequences of the amplified region of ALS gene located in GenBank using a multiple sequence alignment program Clustal Omega.

Note: Several sequences obtained from Source Bioscience were not readable. Therefore, it was necessary to repeat sequencing (Figure 1).

2.2 Multiplex PCR-based analysis of microsatellites in three weedy sunflower populations

Multiplex PCR-based analysis of microsatellites is suitable for studies population variability of weeds. Namely, variation in satellite DNA sequences (different size repeated DNA sequences) can be used to determine genetic differences between organisms or closely related individuals (e.g. weedy sunflower which is result of hybridization between different sunflower forms including crop plant, off-type plants, wild plants, volunteer plants and weedy forms). Satellite loci can be defined by the length of the core repeat, number of repeats or the overall repeat length. Microsatellites are DNA repeats with 2–6 nucleotides in length and they are also called simple sequence repeats (SSRs) or short tandem repeats (STRs).

2.2.1 Plant material

Seeds of three different population of weedy sunflower were collected and sown in the greenhouse. Seedlings were transplanted into larger pots. Fresh leaf material from 10 randomly selected plants from each population was collected for DNA extraction. Some leaf samples (taken from a single plant) were used for DNA extraction immediately after sampling, while some them stored in a freezer (–20°C) until analysis. Before analysis samples were lyophilized after storage at the –80°C during 24h.

2.2.2 DNA extraction

DNA was isolated from about 100 mg of fresh plant leaves according to the QiagenDneasy® Plant Mini Kit following the manufacturer's protocol (https://www.qiagen.com/dz/shop/sample-technologies/dna/dna-preparation/dneasy-plant-mini-kit/). The quality and concentration of extracted DNA were determined spectrophotometrically using a Nanodrop® 1000. DNA extracts were stored at –20°C when not in use.

Note: Yield of DNA from lyophilized samples was low and curve on Nanodrop® 1000 was unacceptable. Possible reason was high concentration of proteins. To avoid that, we tried to add polyvinylpyrrolidone (PVP), but the problem persisted. Extraction was repeated using fresh leaf samples and satisfactory yield of DNA was obtained.

2.2.3 Primers selection

Seven microsatellite loci were selected from Garayalde et al. (2011) and Muller et al. (2010). The SSR flouorescently labelled markers were sorted by allele-length range (Table 1).

Marker Name	Forward primer sequence	Reverse primer sequence	Allele size range
ORS297	FAM-GTGTCTGCACGAACTGTGGT	TGCAAAGCTCACACTAACCTG	214–237
ORS309	FAM-CATTTGGATGGAGCCACTTT	GATGAAGATGGGGAATTTGTG	116–130
ORS337	FAM-TTGGTTCATTCATCCTTGGTC	GGGTTGGTGGTTAATTCGTC	165–197
ORS342	NED-TGTTCATCAGGTTTGTCTCCA	CACCAGCATAGCCATTCAAA	305–361
ORS371	HEX-GGTGCCTTCTCTTCCTTGTG	CACACCACCAAACATCAACC	234–264
ORS432	HEX-TGGACCAGTCGTAATCTTTGC	AAACGCATGCAAATGAGGAT	155–167
ORS656	NED-TCGTGGTAAGGGAAGACAACA	ACGGACGTAGAGTGGTGGAG	181–254

Table 1: Microsatellite loci, fluorescent dye, sequence, allele size range for 7 SSR markers.

The amplification reaction were examined for each primer separately consisted of 0.1 μl of reverse primer and 0.1 μl of forward primer, flouorescently labelled with NED, HEX or FAM, 5 μl MMx2 (Taq), 3.8 μl Rnase-free water and 1μl template DNA in a total volume of 10 μl. Also, the actual amplification reaction for 7 primer together consisted of each unlabelled reserve primer (7 × 0.1 μl) and each of forward primer (7 × 0.1μl), 5 μl MM × 2 (Taq), 2.6 μl H_2O and 1μl template DNA a total volume of 10 μl.

2.2.4 PCR analysis

PCR was done following the manufacturer's protocol Type-it®Microsatellite PCR Handbook (https://www.qiagen.com/dz/resources/search-resources).

Thermal Cycler (Applied Byosistems Verite 95 Well) was programmed for initial denaturation step of 94°C for 5 min, followed by 6 touchdown cycles of 94°C for 30 s, touchdown annealing temperature (Tx) for 90 s (Tx is initially 63°C and decreases of 1°C per cycle for the six first cycles, until it reaches 57°C) and 72°C for 60 s. PCR products were subsequently amplified for 29 cycles at 94°C for 30 s, touchdown annealing temperature 57°C for 90 s and 72°C for 60 s with a final extension at 63°C for 30 min. DNA Fragment Analysis by the Capillary Electrophoresis system is done in Source Bioscience (Nottingham, UK).

Data analysis

GENEMAPPER (Applied biosystem) and PEAK SCANNER software were used for analyses of the DNA fragments and to score the genotypes (Figure 2).

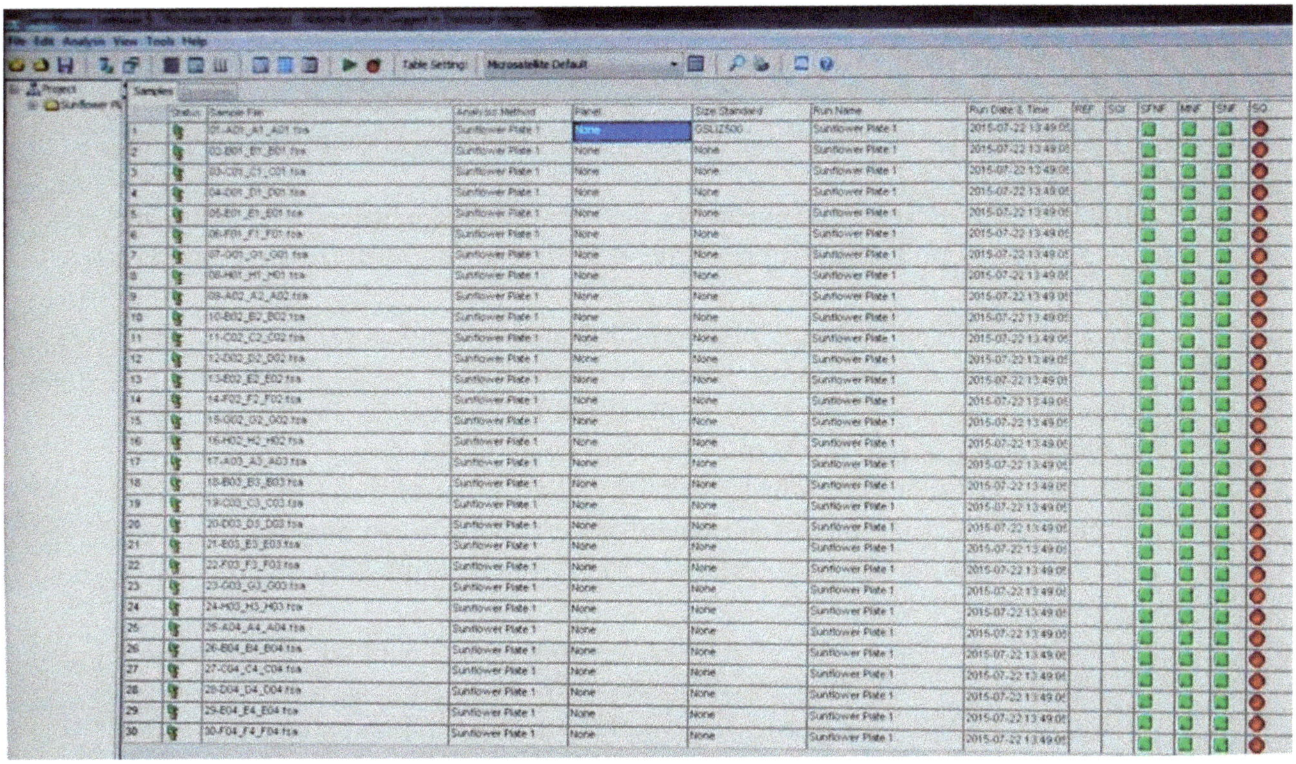

Figure 2: Analyses of the DNA fragments using GENEMAPPER.

3 Acknowledgements

This work was funded by EU Commission project AREA, no. 316004. Authors wish to acknowledge prof. Radmila Stikić who has enabled them to train for molecular research and Dr George Gibbins, senior laboratory technician, for realization of training in School of Agriculture, Policy and Development at University of Reading. Also, we thank Dr TijanaBlanuša for support regarding training realization.

4 References

Garayalde, A.F., Poverene, M., Cantamutto, M. & Carrera, A.D. (2011). Wild sunflower diversity in Argentina revealed by ISSR and SSR markers: an approach for conservation and breeding. *Annals of Applied Biology, 158*, 305–317. DOI: https://doi.org/10.1111/j.1744-7348.2011.00465.x

Gaskin, J. F., Bon, M. C., Cock, M. J., Cristofaro, M., De Biase, A., De Clerck-Floate, R., Ellison, C.A., Hinz, H.L., Hufbauer, R.A., Julien, M.H. & Sforza, R. (2011). Applying molecular-based approaches to classical biological control of weeds. *Biological Control, 58*, 1–21. DOI: https://doi.org/10.1016/j.biocontrol.2011.03.015

Goolsby, J.A., De Barro, P.J., Makinson, J.R., Pemberton, R.W., Hartley, D.M., & Frohlich, D.R. (2006). Matching the origin of an invasive weed for selection of a herbivore haplotype for a biological control programme. *Molecular Ecology, 15*, 287–297. DOI: https://doi.org/10.1111/j.1365-294X.2005.02788.x

NBS Biologicals. (2015). Spin Column Purification DNA Cleanup Handbook.

Retrieved from http://www.nbsbio.co.uk/downloads/DNA_Cleanup_Handbook.pdf

Qiagen. (2015). DNeasy® Plant Handbook. Retrieved from https://www.qiagen.com/dz/shop/sample-technologies/dna/dna-preparation/dneasy-plant-mini-kit/

Qiagen. (2009). Type-it®Microsatellite PCR Handbook. Retrieved from https://www.qiagen.com/dz/resources/search-resources.

Kolkman, J.M., Slabaugh, M.B., Bruniard, J.M., Berry, S., Bushman, B.S., Olungu, C., Maes, N., Abratti, G., Zambelli, A., Miller, J.F., Leon, A. & Knapp, S.J. (2004). Acetohydroxyacid synthase mutations conferring resistance to imidazolinone or sulfonylurea herbicides in sunflower. *Theoretical and Applied Genetics, 109*, 1147–1159. DOI: https://doi.org/10.1007/s00122-004-1716-7

Madeira, P.T., Coetzee, J.A., Center, T.D., White, E.E. & Tipping, P.W. (2007). The origin of Hydrillaverticillata recently discovered at a South African dam. *Aquatic Botany, 87*, 176–180. DOI: https://doi.org/10.1016/j.aquabot.2007.04.008

Michitte, P., De Prado, R., Espinoza, N., Ruiz-Santaella, J.P. & Gauvrit, C. (2007). Mechanism of Resistance to glyphosate in Ryegrass (*Lolium multiflorum*) Biotype from Chile. *Weed Science, 55*, 435–440. DOI: http://dx.doi.org/10.1614/WS-06-167.1

Moss, S.R., Perryman, S.A.M. & Tatnell, L.V. (2007). Managing Herbicide-Resistant Blackgrass (*Alopecurus myosuroides*): Theory and Practice. *Weed Technology, 21*, 300–309. DOI: https://doi.org/10.1614/WT-06-087.1

Muller, M-H., Latreille, M. & Tollon, C. (2010). The origin and evolution of a recent agricultural weed: population genetic diversity of weedy population od sunflower (*Helianthus annuus* L.) in Spain and France. Evolutionary application, *Blackwell, 4*, 499–514. DOI: https://doi.org/10.1111/j.1752-4571.2010.00163.x

Paterson, I. D., Douglas A. D. & Hill, M. P. (2009). Using molecular methods to determine the origin of weed populations of *Pereskia aculeata* in South Africa and its relevance to biological control. *Biological Control, 48*, 84–91. DOI: https://doi.org/10.1016/j.biocontrol.2008.09.012

Slotta T.A.B., Rothhouse J.M., Horvath D.P. & Foley M.E. (2006). Genetic diversity of Canada thistle (*Cirsium arvense*) in North Dakota. *Weed Science 54*, 1080–1085. DOI: http://dx.doi.org/10.1614/WS-06-038R1.1

White, A.D., Graham, M.A. & Owen, M.D.K. (2003). Isolation of acetolactate synthase homologs in common sunflower. *Weed Science, 51*, 845–853. DOI: https://doi.org/10.1614/P2002-136

DNA Extraction and Application of SSR Markers in Genetic Identification of Grapevine Cultivars

Zorica Ranković-Vasić, Dragan Nikolić

Abstract

Microsatellite markers (SSR markers) are widely used in grapevine genetic research for identification of cultivars, parentage analysis, and genetic characterization of germplasm. Aim of this work was extraction of total DNA, primer selection and design, PCR protocols and analysis of DNA sequences with special emphasize on variability between collected samples of different grapevine cultivars. The material used in this study were samples of grapevine leaves of different autochthonous and introduced cultivars from grapevine collection on Experimental field "Radmilovac" at the Faculty of Agriculture, University in Belgrade and from the National fruit collection "Brogdale" from UK. Standard set of nine primers for grapevine was used. Analyses were performed in Molecular Genetics Laboratory, School of Agriculture, Policy and Development, University of Reading, Reading, UK. Extraction and purification of total DNA from fresh and frozen plant material (grapevine leaves) was performed using a DNeasy® Plant Mini (Qiagen Inc.) kit. The concentration of extracted DNA was measured by NanoDrop spectrophotometer and stored on –20°C until use. In the study, we utilized the protocol for Type-it Microsatellite

How to cite this book chapter:
Ranković-Vasić, Z. and Nikolić, D. 2019. DNA Extraction and Application of SSR Markers in Genetic Identification of Grapevine Cultivars. In: Vucelić Radović, B., Lazić, D. and Nikšić, M. (eds.) *Application of Molecular Methods and Raman Microscopy/Spectroscopy in Agricultural Sciences and Food Technology*, Pp. 23–43. London: Ubiquity Press. DOI: https://doi.org/10.5334/bbj.c. License: CC-BY 4.0

PCR Kit, optimized for fluorescent primers, and subsequent high-resolution fragment analysis by capillary sequencing instruments, following the Type-it Microsatellite PCR Handbook (Qiagen Inc.). The results of DNA analyses should be combined with ampelographic descriptors in identification of cultivars and planning the selection of grapevine varieties with desirable viticultural and enological values.

1 Introduction

Grapevine (*Vitis vinifera* L.) is one of the most valuable horticultural species. Currently, there are a large but imprecise number of grapevine cultivars in the world. In many regions have the synonyms (different names for the same cultivar) as well as homonyms (different cultivars identified under the same name). This number could likely be reduced once all cultivars are properly genotyped and compared. Identification of grapevine cultivars based on morphological differences between plants may be incorrect due to the influence of ecological factors. Therefore, methods for analysis at the cultivar genotype level have been developed. In the last twenty years, various techniques for the characterization of cultivars at the level of DNA (RFLP, RAPD, AFLP, SCAR and SSR markers) and isoenzymes have been established. The most appropriate for genotyping are those, using microsatellite markers (Jakše et al. 2013). In the past decade, the application of methods for molecular characterization has been significantly enhanced, particularly, DNA technology in ampelography, helping to identify varieties and their origin. Microsatellite markers (SSR markers) are widely used in grapevine genetic research for identification of cultivars, parentage analysis, and genetic characterization of germplasm. Microsatellites or simple sequence repeats (SSRs) have proved to be the most effective markers for grapevine genotyping (Sanchez-Escribano et al. 1999; Laucou et al. 2011). Thomas and Scott (1993) first used microsatellites for the identification of grapevine cultivars and demonstrated that microsatellite sequences are often represented in the grapevine genome and are very informative for the identification of *Vitis vinifera* cultivars. Hundreds of microsatellite markers for grapevines have been developed and most of them are publicly available (Bowers et al. 1996; Arroyo-Garcia & Martinez-Zapater, 2004; Adam-Blondon et al. 2004; Merdinoglu et al. 2005; Cipriani et al. 2008). A set of six (VVS2, VVMD5, VVMD7, VVMD27, VrZag62, VrZAG79) or nine (previous six, combined with the following three: VVMD32, VVMD36, VVMD25) microsatellite markers has been used in grapevine genotyping studies, mostly for determining genetic variability among European grapevine cultivars, which are highly polymorphic (Sefc et al. 2001; This et al. 2004; Žulj et al. 2013). Aim of this research was extraction total DNA, primer selection and design, PCR protocols and analysis of DNA sequences with special emphasize on variability between collected samples of different grapevine cultivars.

2 Materials, Methods and Notes

2.1 Plant material

The material used in this study, were the samples of grapevine leaves of different grapevine cultivars. The source of the material was either the developed leaves from vines in the vineyard from collection "Brogdale", UK (leaves should be the size of a few centimeters, Fig. 1a, b, c), and leaves obtained from cuttings in the laboratory (the method of "provocation"), from collection "Radmilovac", Serbia (Fig. 2a, b).

Note:

- You can not use partially developed or fully developed buds (Fig. 3).
- Buds have a high concentration of protein.
- Isolation of DNA will fail (will be very difficult) if extraction is carried out from the buds (if used Kit); would not provide adequate DNA concentration.

Figure 1: Collection of plant material from the vineyard.

Figure 2: Plant material obtained from cuttings in the laboratory.

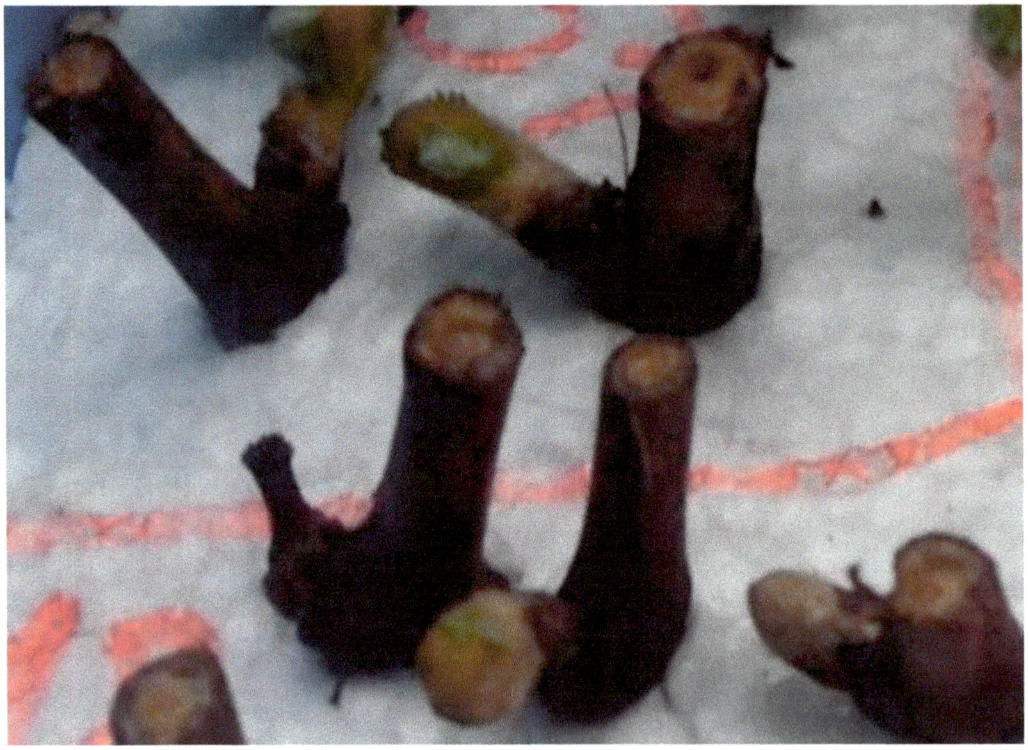

Figure 3: Buds.

Figure 4: Leaves in the freezer.

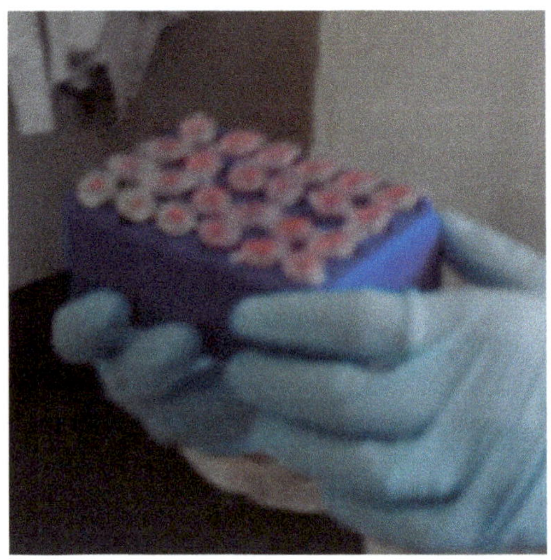

Figure 5: Tubes for sample keeping .

- The leaves can be kept in the freezer (in paper bags) until the beginning of DNA isolation (Fig. 4)
- The samples can be kept in the 1.5 or 2 ml tubes (–20°C) (Fig. 5).
- The samples can be lyophilized (weight about 20 mg), but the extracted DNK is not of desirable quality (Fig. 6).

Figure 6: Measurement of the sample weight.

- Working with lyophilized samples is more difficult (weight measurement of samples is complicated).
- Each sample must have a code.

2.2 DNA extraction

Extraction and purification of total DNA from fresh or frozen plant material (grapevine leaves) was performed using a DNeasy ® Plant Mini Kit following the standard protocol for isolation of DNA from plant leaf tissue outlined in the DNeasy Plant protocol handbook (Qiagen Inc.).

Notes before starting:

- Perform all centrifugation steps at room temperature (15–20°C).
- If necessary, redissolve any precipitates in buffer AP1 and buffer AP3/E concentrates.
- Add ethanol to buffer AW and buffer AP3/E concentrates.
- Preheat a water bath or heating block to 65°C.

Extraction protocol:

1. Plant leaves (about 150–170 mg fresh material) (Fig. 7) are grinded under liquid nitrogen (Fig. 8) to a fine powder using a mortar and pestle (Fig. 9) or Tissue Lyser (Fig. 10). The tissue powder and liquid nitrogen were transferred to 1.5 ml tube and allowed the liquid nitrogen to evaporate (Fig. 11).

Figure 7: Measurement of the sample weight.

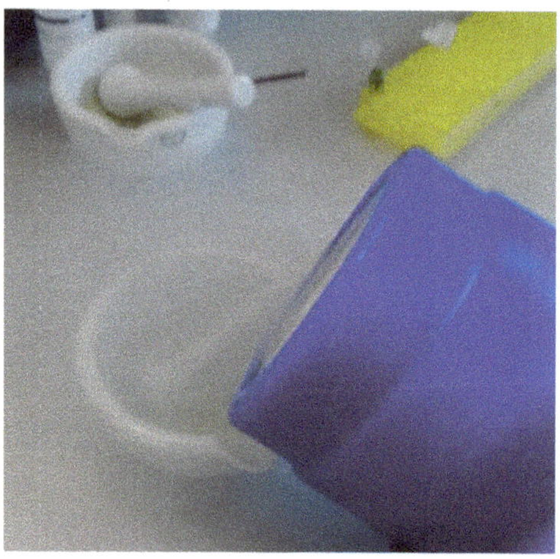

Figure 8 & 9: Grinding under liquid nitrogen.

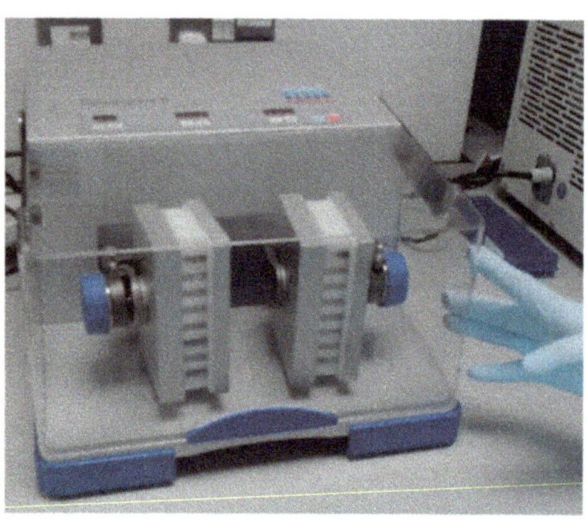

Figure 10: Grinder.

Figure 11: Tubes for samples.

2. 400 μl of Buffer AP1 and 4 μl of RNase were added into the tube(s) with material and vortexed vigorously (Fig. 12).

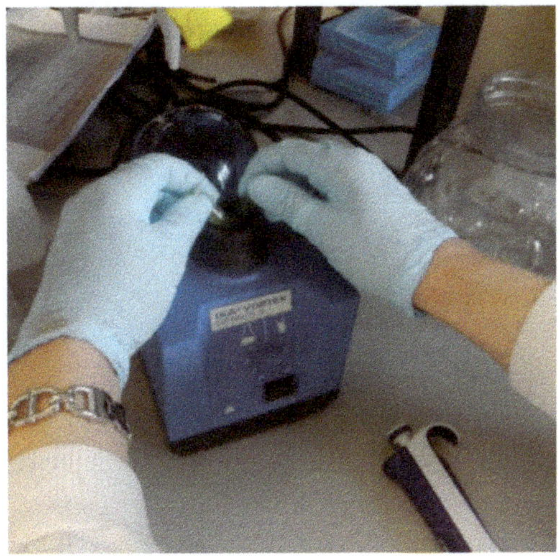

Figure 12: Vortexing.

3. The tubes were incubated at 65°C for 10 min on Termomixer that was set up to shake from 450 to 500 rpm (Fig. 13). During this step cells were lysed.

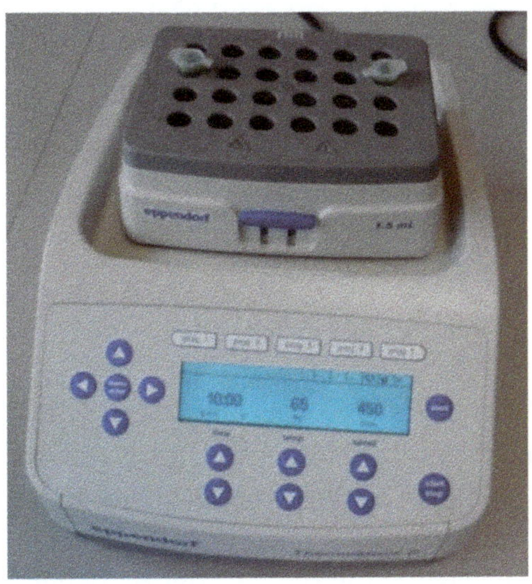

Figure 13: Termomixing.

4. 130 μl of Buffer AP2 (P3) were added to the lysate, mixed, and tubes were incubated for 5 min on ice. During this step detergent, proteins and polysaccharides were precipitated. The tubes with lysate were centrifuged for 5 min at 14 000 rpm speed (or on max 14 680 rpm) in order to remove the precipitates (Fig. 14). The samples in the centrifuge must be uniformly distributed (in equilibrium) (Fig. 15).

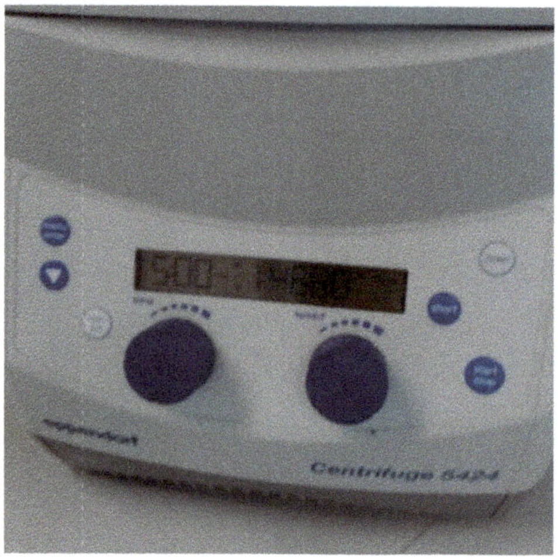

Figure 14: Centifugation.

Figure 15: Position of the samples in centrifuge.

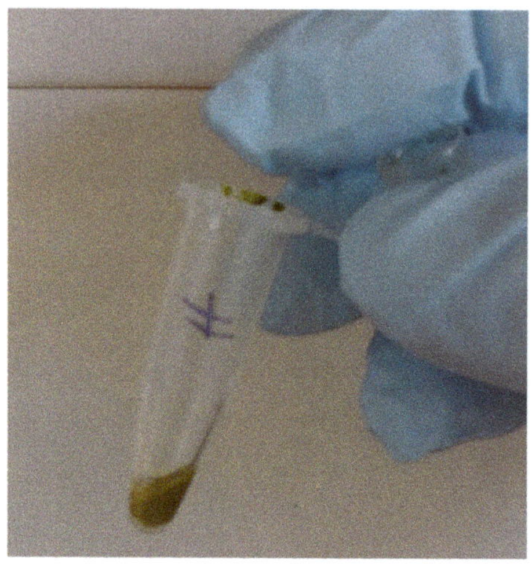

Figure 16: Lysate.

5. The obtained lysate (Fig. 16) was applied to the QIAshredder spin column placed in a 2 ml collection tube (with pink cover) and centrifuged for 2 min at 14 000 rpm speed (or max speed on 14 680 rpm).
6. The flow-through fraction (400–450 μl) from step 5 was transferred to a new tube without disturbing the cell-debris pellet.
7. 600 μl of Buffer AW1 were added to the cleared lysate and mixed by pipetting.
8. 650 μl of the mixture from step 7 were transferred to the DNeasy mini spin column sitting in a 2 ml collection tube and centrifuged for 1 min at 8000 rpm. After that flow-through were discarded.
9. The centrifugation at 8 000 rpm for 1 min was repeated.
10. DNeasy column were placed in a new 2 ml collection tube and 500 μl Buffer AW2 (AW) were added to the DNeasy column and centrifuged for 1 min at 8 000 rpm. After that flow-through were discarded.

11. 500 µl Buffer AW2 (AW) were added to the DNeasy column and centrifuged for 2 min at 14 000 rpm (or max 14680 rpm) to dry the membrane. It is important to dry the membrane of the DNeasy column since residual ethanol may interfere with subsequent reactions. This spin ensures that no residual ethanol will be carried over during elution. After centrifugation flow-through were discarded.
12. The DNeasy column were transferred to a 1.5 ml tubes and pipeted 60 µl of preheated (65°C) Buffer AE directly onto the DNeasy membrane. The column were incubated for 5 min (may be up to 15 minutes) at room temperature and then centrifuged for 1 min at 8 000 rpm to elute DNA.
13. Step 12 was repeated.
14. The filter is removed. Tube with the DNA sample should be closed and placed on ice.

Note:

- Always mark the tubes used in the extraction.
- Take care of the cleanliness of the desk.
- The working surface has to be cleansed several times during the procedure
- Always wear clean gloves (change the gloves several times during the work).
- For each pipetting must be put new sequel to the pipette.

2.3 Measuring the DNA concentration

Spectrophotometry is used to determine the concentration of DNA in the sample. The concentration of extracted DNA was measured by NanoDrop spectrophotometer (Fig. 17) and storage on –20°C until use.

Procedure:

1. Turn on the laptop.
2. On the screen click NanoDrop 2000.
3. Click on Nucleic Acid.
4. 1.5 µl of blank Buffer AE were added on the needle spectrophotometer.
5. Spectrophotometers closed.
6. Click on the blank.
7. Wait for sensing.
8. Wipe the spectrophotometer needle with a paper towel.
9. 1.5 µl DNA sample from ice were added on the spectrophotometer needle.
10. Click on measure. Typical DNA Spectrum is shown in the Fig. 18a and the one obtained during measuring of extracted DNA from our samples is shown in Fig. 18b.

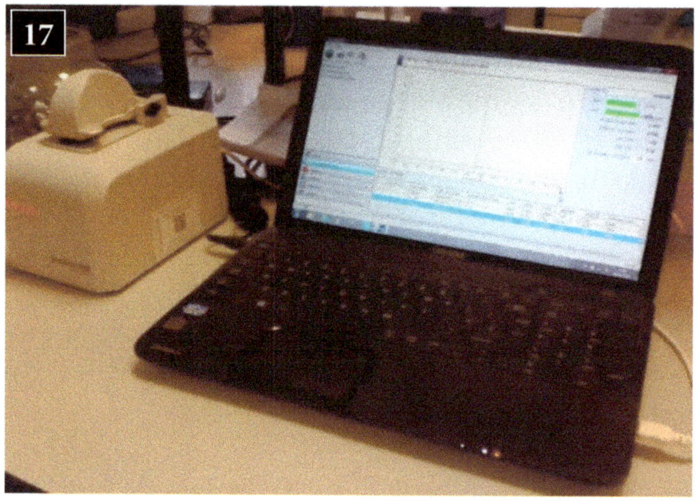

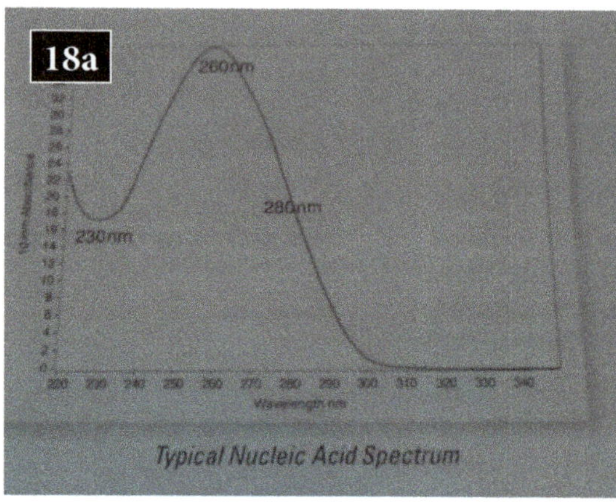

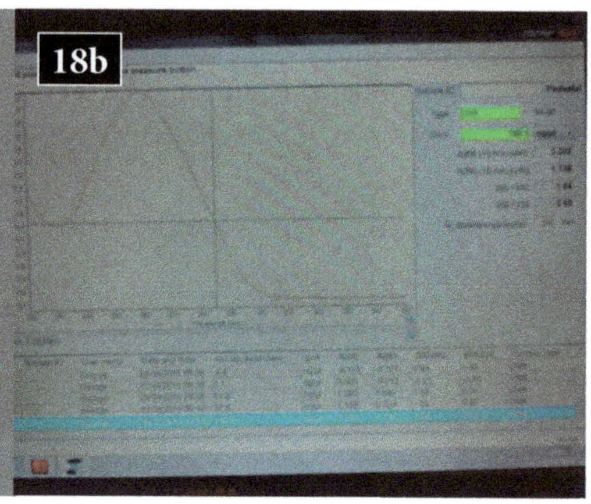

Figures 17, 18a and 18b: Measuring the DNA concentration.

Code	Conc. DNA (ng/µl)	A260	A280	260/280	260/230
28	175.6	3.513	1.907	1.84	2.66
31	128.3	2.566	1.439	1.78	2.05
32	335.8	6.716	3.660	1.84	2.60
34	89.3	1.786	0.975	1.83	2.76

Table 1: Measured concentration of DNA in different grapevine cultivars.

Note:

- When configuring, spectrophotometer must be closed.
- When the blank is read, should not get any spectrum on the screen (without spectrum).
- Good concentration of DNA is shown in Table 1.

- The ration of 260/280 and 260/230 must be higher of 1.7, for extracted DNA, to be considered as high-quality material, suitable for further analyses.
- In addition to estimate the quality of purified total DNA by calculating the ration 260/280 and 260/230, we also analyzed all samples by gel electrophoresis (see 2.6.1).

2.4 PCR amplifications

2.4.1 Microsatellite analysis

Analysis was performed using nine microsatellite loci: VVS2 (Thomas and Scott, 1993), VVMD5, VVMD7, VVMD25, VVMD27, VVMD28, VVMD32 (Bowers et al., 1996, 1999), VrZAG62 и VrZAG79 (Sefc et al., 1999). This set of highly polymorphic markers was used by the European Grape-Gen06 consortium (http://www1.montpellier.inra.fr/grapegen06/accueil.php) as the standard set for the screening of grapevine collections. Equivalent list primers for *Vitis vinifera* L. is shown in Table 2. All forward primers were labeled with 6-FAM, VIC, PET, or NED fluorescent dyes.

Note:
The DNA analyses should be combined with ampelographic descriptions, IPGRI, UPOV, OIV (1997) in planning the selection of grapevine varieties with desirable viticultural and enological value.

Material needed for PCR amplification:

1. Taq enzyme mix
2. Primers (R and F)
3. Samples
4. Deionized water

Master mix (for 1 sample):

1. Taqenzymemix (10×1 μl)
2. Primers R (0.7 μl)
3. Primers F (0.7 μl)
4. Deionized water (7.6 μl)

Note:

- Total of master mix is 19 μl.
- All this is carefully shaken with pipette about 20 times.
- A volume of master mix components is multiplied by the number of samples (for bigger number of samples).

Name		Sequence	Source	Fluorescent
VVS2	„forward”	**5'-CAG CCC GTA AAT GTA TCC ATC-3'**	Thomas and Scott (1993.)	**6FAM**
	„reverse”	**5'-AAA TTC AAA ATT CTA ATT CAA CTG G-3'**		
VVMD5	„forward”	**5'-CTA GAG CTA CGC CAA TCC AA-3'**	Bowers et al. (1996.)	**6FAM**
	„reverse”	**5'-TAT ACC AAA AAT CAT ATT CCT AAA-3'**		
VVMD7	„forward”	**5'- AGA GTT GCG GAG AAC AGG AT -3'**	Bowers et al. (1999.)	**VIC**
	„reverse”	**5'- CGA ACC TTC ACA CGC TTG AT -3'**		
VVMD25	„forward”	**5'-TTC CGT TAA AGC AAA AGA AAA AGG-3'**	Bowers et al. (1999.)	**NED**
	„reverse”	**5'-TTG GAT TTG AAA TTT ATT GAG GGG-3'**		
VVMD27	„forward”	**5'-GTA CCA GAT CTG AAT ACA TCC GTA AGT-3'**	Bowers et al. (1999.)	**NED**
	„reverse”	**5'-ACG GGT ATA GAG CAA ACG GTG T-3'**		
VVMD28	„forward”	**5'-AAC AAT TCA ATG AAA AGA GAG AGA GAG A-3'**	Bowers et al. (1999.)	**6FAM**
	„reverse”	**5'-TCA TCA ATT TCG TAT CTC TAT TTG CTG-3'**		
VVMD32	„forward”	**5'-TAT GAT TTT TTA GGG GGG TGA GG-3'**	Bowers et al. (1999.)	**PET**
	„reverse”	**5'-GGA AAG ATG GGA TGA CTC GC-3'**		
VrZAG62	„forward”	**5'-GGT GAA ATG GGC ACC GAA CAC ACG C-3'**	Sefc et al. (1999.)	**PET**
	„reverse”	**5'-CCA TGT CTC TCC TCA GCT TCT CAG C-3'**		
VrZAG79	„forward”	**5'-AGA TTG TGG AGG AGG GAA CAA ACC G-3'**	Sefc et al. (1999.)	**NED**
	„reverse”	**5'-TGC CCC CAT TTT CAA ACT CCC TTC C-3'**		

Table 2: Equivalent list primers on *Vitis vinifera* L.

Procedure:

1. In the PCR tube add 1 μl isolated DNA and 19 μl master mix.
2. In the control PCR tube add 1 μl deionized water and 19 μl master mix.
3. Centrifuge for 1–2 sec.
4. Sett of PCR machine – Veriti™ Thermal Cycler (Applied Biosystems, Foster City, California, USA) (Fig. 19).

Note:

- PCR machine typically works about 1.30 hours.
- Depending on primers and overall experimental design, the different temperature and time settings could be used (Fig. 20). Different temperature is shown in Table 3.
- PCR reactions in a Veriti™ Thermal Cycler (Applied Biosystems, Foster City, California, USA) using the following conditions: 94°C for 2 min, 35 cycles of 1 min at 94°C, 1 min at 50°C, and 1 min at 72°C, with a final extension of 30 min at 72°C.
- PCR reactions in a GeneAmp PCR 9700 thermocycler (Applied Biosystems, Foster City, CA) using the following conditions: 94°C for 5 min, 30 cycles of 1 min at 94°C, 1 min at 51°C or 49°C (for multiplex or singleplex PCR, respectively), and 1 min at 72°C, with a final extension of 30 min at 72°C.
- Can be used singleplex and multiplex reactions. Two multiplex PCR reactions were carried out for five (VVS2, VVMD7, VVMD27, VrZAG62, VrZAG79) and three (VVMD25, VVMD28, VVMD32) of the analyzed SSR a singleplex for VVMD5 (Vilanova et al. 2009).

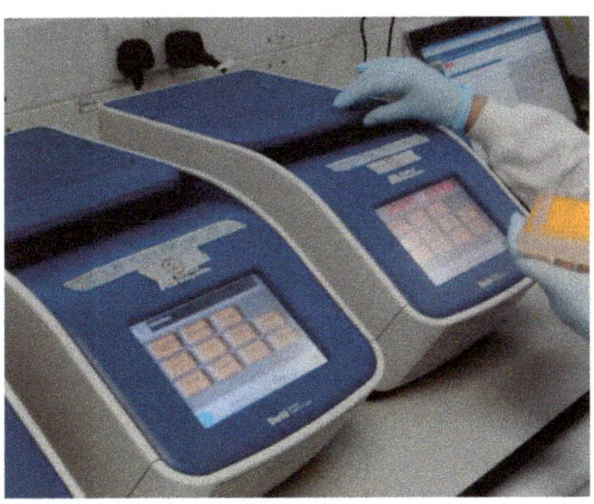

Figure 19: PCR mashine.

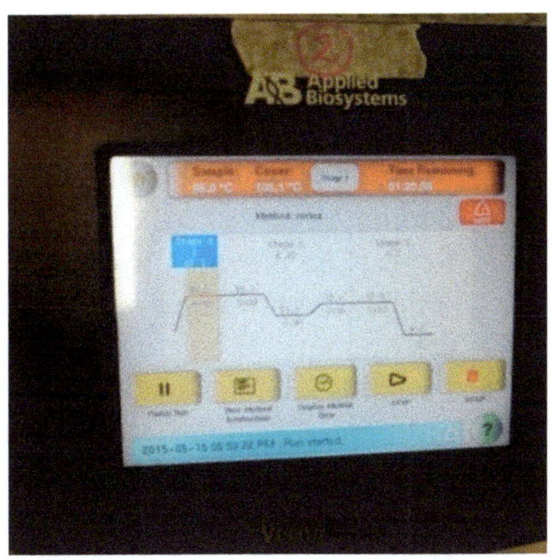

Figure 20: PCR mashine settings.

PCR amplification steps	Primer				
	VVS2, VVMD5, VrZAG79, VVMD7	**VVMD27 VVMD28**	**VrZAG62**	**VVMD25 VVMD32**	
Precycle/ Initialization	95°C 1 min	94°C 2 min	94°C 2 min	94°C 2 min	
Denaturation	95°C 30 sec	92°C 40 sec	92°C 40 sec	92°C 40 sec	30x or 35x
Annealing	51°C 30 sec (57–58°C, 30 sec) (48–56°C, 30 sec)	55–56°C 30 sec 49°C 30 sec	56–59°C 40 sec	44–45°C 30 sec 48–49°C 30 sec	
Elongation/ Extension	72°C 30 sec	72°C 2 min	72°C 2 min	72°C 2 min	
Postcycle/Final elongation	72°C 1 min	72°C 5 min	72°C 5 min	72°C 5 min	

Table 3: Suggestions of different temperature for different primers in PCR.

2.5 Preparing for sequencing analysis (Protocol for Purification of PCR Products):

Transfer PCR reaction mixture (whole quantity, about 35 μl) to a 1.5 ml microfuge tube (blue cover) and add 3 volumes of Binding buffer 1 (105 μl). Then, according to the protocol described in DNA Cleanup Handbook at http://www.nbsbio.co.uk/downloads/DNA_Cleanup_Handbook.pdf

1. Transfer the above mixture solution to the spin column and keep it at room temperature for 2 minutes. Centrifuge at 10 000 rpm for 2 minutes. Discharge the eluate from the tube.
2. Add 750 μl of Wash solution to the column and spin at 10 000 rpm for 2 minutes.
3. Repeat the previous washing procedure using the same conditions. In order to remove any residual wash solution, extend spinning duration for 1 minute.
4. Place the column into a clean 1.5 ml tube and add 30–50 μl (usually 30 μl) of Elution buffer exactly into center of the column. Leave it at room temperature for 2 minutes (can also stand for 5 minutes). Centrifuge at 10 000 rpm for 2 minutes to elute the DNA.
   Note:
   - The incubation of the column, with the Elution buffer, at higher temperatures (up to 50°C) may slightly increase the yield especially for large DNA.

- The purified DNA can be stored at 4°C for immediate use or in a deep freezer at –20°C for use in future.
- It is recommended to use the DNA Gel Extraction Kit if non-specific amplified DNA fragments are formed in a PCR experiment.

2.6 *Electrophoresis*

Electrophoresis is a method that is used to separate DNA fragments and determine their sizes by comparing them to the sizes of known fragment lengths. In our protocol, we applied this meted to test purified total DNA (2.2.) and the PCR products (2.4.)

2.6.1 Gel Electrophoresis

Procedure for making agarose gel:

1. Measure 0.25 g of the agarose and 25 ml × 1 of TAE buffer (100 ml TAE × 10 add 1000 ml distilled water).
   Note: Agarose gels: Commonly are used concentrations of agarose gel from 0.7% to 2% depending on the size of bands needed to be separated. Measure 0.60 g of the agarose and 60 ml × 1 of TAE buffer for biggest size electrophoresis box. Simply adjust the amount of starting agarose to %g/100 mL TAE (i.e. 2g/100mL will give you 2%).
2. Dissolve agarose by heating in microwave oven (20–30 sec).
   Note: Caution **HOT!** Be careful stirring, eruptive boiling can occur.
3. Let agarose solution cool down for 5 min.
4. According to https://www.addgene.org/protocols/gel-electrophoresis/ add ethidium bromide (EtBr) to a final concentration of approximately 1.0 μg/ml. Ethidium bromide binds to the DNA and allows visualization of the DNA when the gel is exposed to ultraviolet (UV) light.
   Note:
   - **EXTREME CAUTION!** It is known that ethidium bromide is a mutagen. It is necceassary to wear gloves, a laboratory coat and safety glasses when using this techique.
   - Mildori Green Nucleic Acid Staning Solution is a safe alternative to traditional EtBr stain for detecting DNA in agarose gels.
5. Position the well comb in place and pour the agarose into a gel tray.
   Note: Avoid formation of air bubbles which will disrupt the gel network by pouring agarose solution slowly. If bubbles are formed use a pipette tip to push them away towards the edges of the gel.
6. Cool newly poured gel for 10–15 minutes in a refrigarator or let it cool at room temperature for 20–30 minutes, until it completely solidifies.

Preparation of samples for electrophoresis:

1. Carefully load your samples DNA (4μl) on the plate designated for mixing sample with sample buffer.
2. Add loading buffer to each of your samples DNA (2 μl "stop blue").

Procedure of placing gel in electrophoresis box:

1. Remove protection tape.
2. Put the tray in electrophoresis box.
3. Remove the comb.
4. Fill gel box with 1×TAE until the gel is covered.

Loading samples and running an agarose gel:

1. Carefully load a molecular weight ladder (DNA 1000) into the first lane of the gel (add 3–4 μl).
2. Carefully load your DNA samples (4 μl of DNA + 2 μl of loading buffer – "stop blue") into the additional wells of the gel (Fig. 21).
3. Carefully load a molecular weight ladder (DNA 1000) into the last lane of the gel (add 3–4 μl).
4. Apply power supply at 80–150V to the gel until the dye reaches approximately 25–20% of the way from the end of the gel (Fig. 22).
   Note:
   - Negatively charged DNA molecules will move towards the positive electrode. Always place the gel in such a way that samples migrate towards positive electrode (red is positive electrode – connect it to the end of the gel; black is negative electrode – connect it to the start of the gel) (Fig. 23a, b).
   - A typical run time is about 1–1.5 hours, depending on voltage.
5. Turn OFF power supply, disconnect the electrodes from the power source, and then carefully remove the tray with a gel from the gel box.
   Note:
   - TAE buffer can be replenished and used up to 10 times.
   - Tape is obligatory placed at the gel box (write how many times used buffer) (Fig. 24).
   - For a gel of small to mid-size, good separation of DNA fragments is achieved if gel is running at 50V for about 45 minutes (Fig. 25).

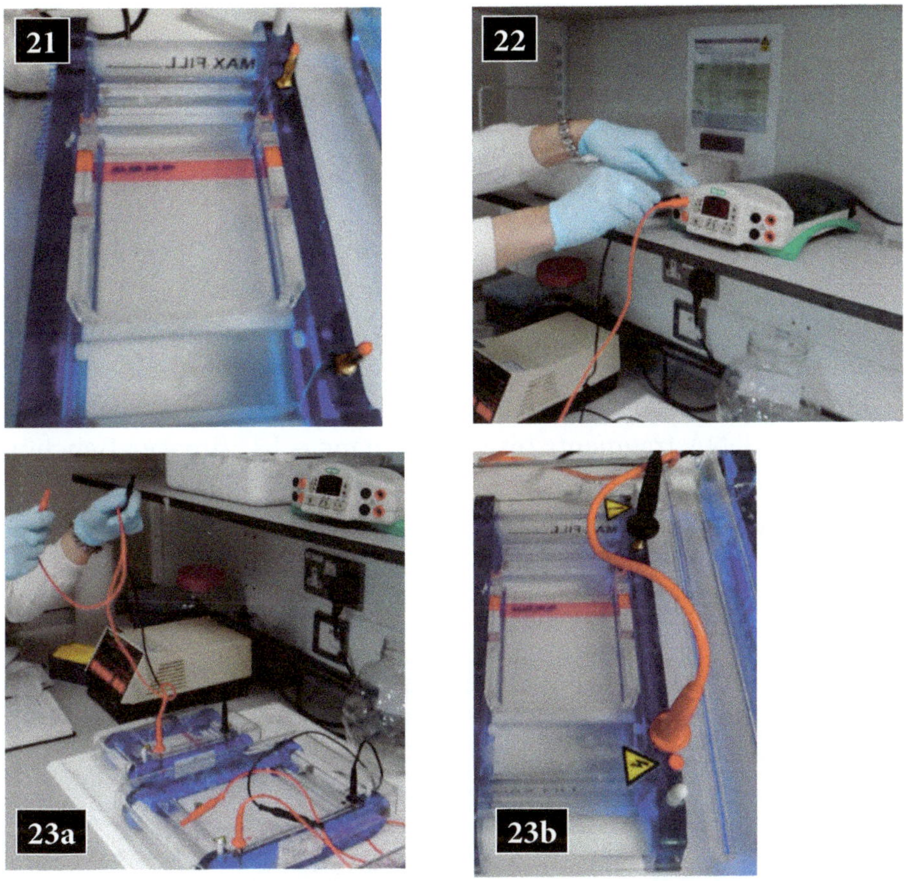

Figures 21., 22, 23a and 23b: Loading samples and running an agarose gel.

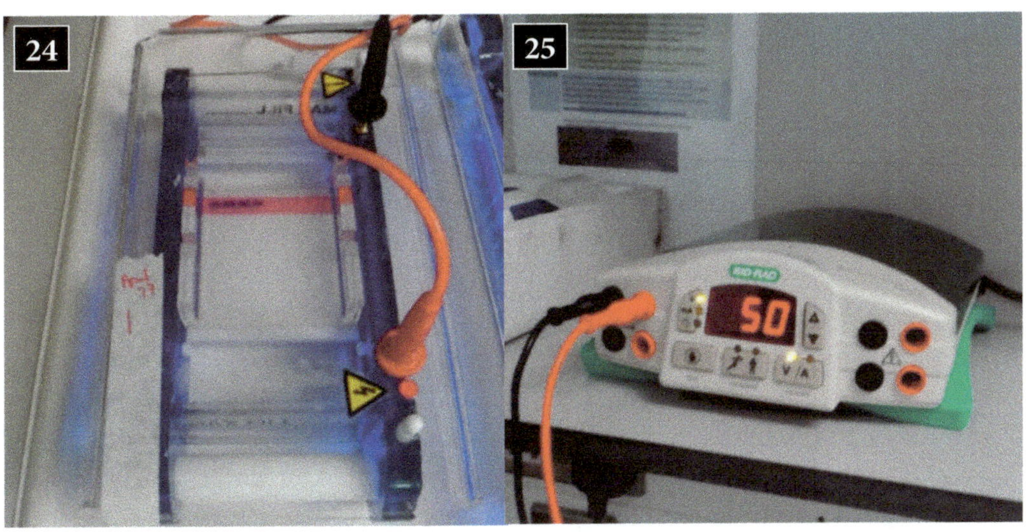

Figures 24. and 25: Electrophoresis apparatus.

2.7 Visualization of DNA fragments

1. By using any device that has a source of UV light your DNA fragments will visualize and look like bands on the gel. According to https://www.addgene.org/protocols/gel-electrophoresis/ pay attention to the following:
   - UV light is very harmfool to your eyes and skin. Wearing protective glasses, gloves and laboratory coat is necessary.
   - Using long-wavelength UV and the shortest possible exposition time will minimize damage to the DNA in the case that further analysis of DNA is planned.
2. Visualize DNA fragments by WP (GelDoc – It^{TS2}/Imager Benchop UV Transilluminator) (Fig. 26).
3. Gel in Transilluminator (Fig. 27).
4. Software: UVP TS2 (Fig. 28).
5. Printing by digital graphic printer UP-D897.

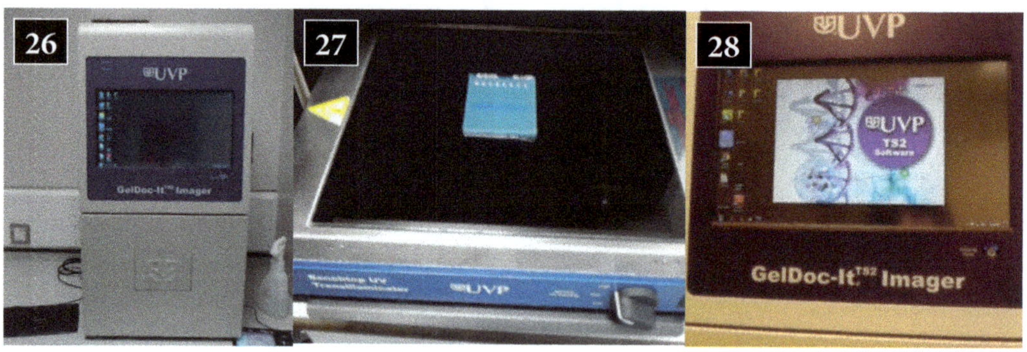

Figures 26., 27. and 28: Visualization of DNA fragments.

2.8 Analyzing Gel:

Use the DNA ladder, in the first lane, as a guide (the manufacturer's instruction will indicate the size of each band) to determine the length of the bands detected in the sample lanes and visualize purified total DNA (Fig. 29) and also visualize the PCR products (Fig. 30).

Note (according to https://www.addgene.org/protocols/gel-electrophoresis/):

- In order to get better resolution (crispness) of samples DNA bands lower voltage in a longer duration of a run can be used. Alternative is to choose a wider gel comb or to load lesser amount of DNA into the well.
- In order to get better separation of bands in the case of similarly sized fragments a higher percentage of agarose gel can be used to better separate smaller bands, and a lower percentage of agarose gel to separate larger bands.

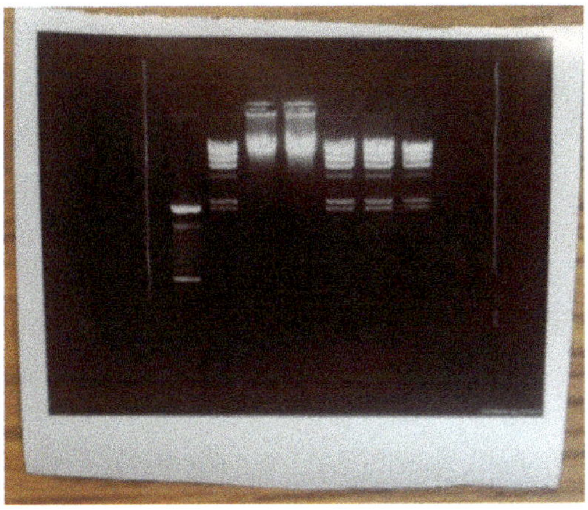

Figure 29: Visualisation of purified total DNA.

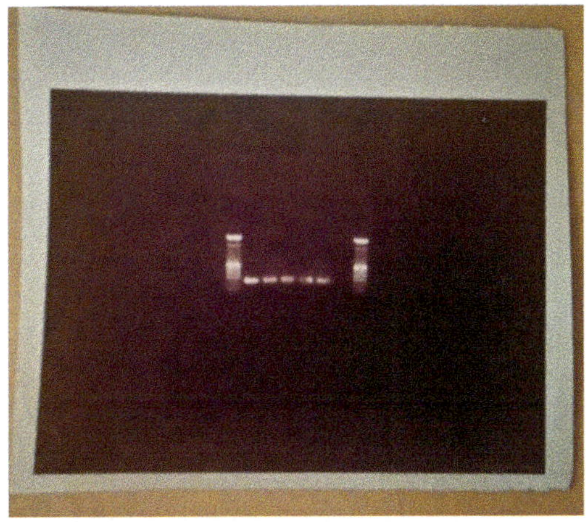

Figure 30: PCR products.

3 Acknowledgements

This work was realized as a part of project: FP7-REGPOT-2012-2013-1. No. 316004-AREA supported by the European Union. Investigations were carried thanks to the help and support dr George Gibbings, dr Matthew Ordidge, dr Tijana Blanusa, and dr Edward Paul Venison, from School of Agriculture, Policy and Development, University of Reading, Reading, UK

4 References

Agarose Gel Electrophoresis, Retrived from https://www.addgene.org/protocols/gel-electrophoresis/

Adam-Blondon, A.F., Roux, C., Claux, D., Butterlin, G., Merdinoglu, D. & This, P. (2004). Mapping 245 SSR markers on the *Vitis vinifera* genome: a tool for grape genetics. *Theoretical and Applied Genetics, 109(5)*,1017–1027. DOI: https://doi.org/10.1007/s00122-004-1704-y

Arroyo-Garcia, R. & Martinez-Zapater, J.M. (2004). Development and characterization of new microsatellite markers for grape. *Vitis, 43(4)*,175–178.

Bowers, J.E., Vignani, R. & Meredith, C.P. (1996). Isolation and characterization of new polymorphic simple sequence repeat loci in grape (*Vitis vinifera* L.). *Genome, 39*, 628–633, DOI. https://doi.org/10.1139/g96-080

Bowers, J.E., Dangl, G.S. & Meredith, C.P. (1999). Development and characterization of additional microsatellite DNA markers for grape. *American Journal of Enology and Viticulture, 50*, 243–246.

Cipriani, G., Marrazzo, M.T., Di Gaspero, G., Pfeiffer, A., Morgante, M. & Testolin, R. (2008). A set of microsatellite markers with long core repeat optimized for grape (Vitis spp.) genotyping. *Bmc Plant Biology, 8*(127), 1–13. DOI: https://doi.org/10.1186/1471-2229-8-127

DNA Cleanup Handbook, Retrived from http://www.nbsbio.co.uk/downloads/DNA_Cleanup_Handbook.pdf

Jakse, J., Štajner, N., Tomić, L. & Javornik, B. (2013). Application of microsatellite markers in grapevine and olives. *Agricultural and Biological Sciences » "The Mediterranean Genetic Code – Grapevine and Olive"*, book edited by Danijela Poljuha and Barbara Sladonja, Intech Open Science, DOI: https://doi.org/10.5772/53411

Laucou, V., Lacombe, T., Dechesne, F., Siret, R., Bruno, J.P., Dessup, M., Dessup, T., Ortigosa, P., Parra, P., Roux, C., Santoni, S., Varès, D., Péros, J.P., Borsiquot, J.M. & This, P. (2011). High throughput analysis of grape genetic diversity as a tool for germplasm collection management. *Theoretical and Applied Genetics, 122*(6),1233–1245. DOI: https://doi.org/10.1007/s00122-010-1527-y

Merdinoglu, D., Butterlin, G., Bevilacqua, L., Chiquet, V., Adam-Blondon, A.F. & Decroocq, S. (2005). Development and characterization of a large set of microsatellite markers in grapevine (*Vitis vinifera* L.) suitable for multiplex PCR. *Molecular Breeding, 15(4)*, 349–366. DOI: https://doi.org/10.1007/s11032-004-7651-0

IPGRI, UPOV, OIV. (1997). Descriptors for Grapevine (Vitis spp.). International Union for the Protection of New Varieties of Plants, Geneva, Switzerland/Office International de la Vigne et du Vin, Paris, France/International Plant Genetic Resources Institute, Rome, Italy.

Sanchez-Escribano, E.M., Martin, J.R., Carreno, J. & Cenis, J.L. (1999). Use of sequence-tagged microsatellite site markers for characterizing table grape cultivars. *Genome, 42(1)*,87–93. DOI. https://doi.org/10.1139/g98-116

Sefc, K.M., Regner, F., Tutetschek, E., Gloessl, J. & Steinkellner, H. (1999). Identification of microsatellite sequences in *Vitis riparia* and their applicability for genotyping of different *Vitis* species. *Genome, 42*, 367–373.

Sefc, K.M., Lefort, F., Grando, M.S., Scott, K., Steinkellner, H. & Thomas, M.R. (2001). Microsatellite markers for grapevine: A state of the art. In: *Molecular Biology and Biotechnology of Grapevine*. Roubelakis-Angelakis KA, editor. Amsterdam: Kluwer Publishers, 407–438. DOI: https://doi.org/10.1007/978-94-017-2308-4

This, P.,Jung, A., Boccacci, P., Borrego, J., Botta, R., Costantinim, L., Crespan, M., Dangl, G.S., Eisenheld, C., Ferreira-Monteiro, F., Grando, S.,Ibáñez, J.,Lacombe, T., Laucou, V., Magalhães, R., Meredith, C.P., Milani, N., Peterlunger, E., Regner, F., Zulini, L. & Maul, E. (2004). Development of a standard set of microsatellite reference alleles for identification of grape cultivars. *Theoretical and Applied Genetics, 109(7)*, 1448–1458. DOI: https://doi.org/10.1007/s00122-004-1760-3

Thomas, M.R. & Scott, N. (1993). Microsatellite repeats in grapevine reveal DNA polymorphisms when analysed as sequence-tagged sites (STSs). Theoretical and *Applied Genetics, 86*, 985–990. DOI: https://doi.org/10.1007/BF00211051

Vilanova, M., Fuente, de la M., Fernández-González, M. & Masa, A. (2009). Identification of New Synonymies in Minority Grapevine Cultivars from Galicia (Spain) Using Microsatellite Analysis. *American Journal of Enology and Viticulture, 60*, 236–240.

Žulj Mihaljević, M., Šimon, S., Pejić, I., Carka, F., Sevo, R., Kojić, A., Gaši, F., Tomić, L., Jovanović Cvetković, T., Maletić, E., Preiner, D., Božinović, Z., Savin, G., Cornea, V., Maraš, V., Tomić Mugoša, M., Botu, M., Popa, A. & Beleski, K. (2013). Molecular characterization of old local grapevine varieties from South East European countries. *Vitis, 52(2)*, 69–76.

Application of qPCR Method for Investigation of Plant Colonization by Human Pathogen Bacteria

Kljujev Igor

Abstract

The consumption of vegetables is very important for prevention cardiovascular diseases and it is recommended by WHO. The fresh vegetables are essential for healthy nutrition and provide minerals and vitamins. The vegetables are mostly consumed raw and it is very important to avoid its microbiological contamination during the production chain. The disease which are caused by human pathogen bacteria are very big problem for public health. These bacteria are able to contaminate fresh vegetables in any part of chain of food production.

The Salmonellosis is the usual foodborne infection which is caused by bacteria *Salmonella spp.* According to U.S. Public Health Service (2009), *Salmonella* is in the second place as causer of foodborne disease in the USA. Approximately, there are 40 000 cases of Salmonellosis in the USA per year (DFBMD 2009).

The most common serovars which are found worldwide are *Salmonella enteritidis* and *Salmonella typhimurium* but others serovars are limited to specific regions in the world (OIE 2005).

How to cite this book chapter:

Igor, K. 2019. Application of qPCR Method for Investigation of Plant Colonization by Human Pathogen Bacteria. In: Vucelić Radović, B., Lazić, D. and Nikšić, M. (eds.) *Application of Molecular Methods and Raman Microscopy/Spectroscopy in Agricultural Sciences and Food Technology*, Pp. 45–57. London: Ubiquity Press. DOI: https://doi.org/10.5334/bbj.d.

Today, there is lot of methods for detection human pathogen bacteria in food. The conventional methods are generally timeconsuming. The PCR is much faster and with qPCR we can get results only in few hours.The PCR allows increasing speed, sensitivity, specificity of detection of human pathogen bacteria in fresh vegetables.

The aim of this study is application of the qPCR method for detection of *Salmonella enteric*a subsp. Welteweden and *Salmonella typhimurium* LT2 in wheat seedlings.

Bacterial strains used in this experiment were *Salmonella enteric*a subsp. Welteweden and *Salmonella typhimurium* LT2. The model plant was wheat. The bacterial suspension applied for inoculation seeds was ≈ 10^8 CFU. The inoculated plants left to grow in phitochamber for 3 weeks. The standard PCR was done for *Salmonella* strains. The primers for *Salmonella* were: rfbJ; fliC; fijB; invA, hilA. It was done cloning for getting plasmid with invA gene which was used for preparing standards for qPCR. Isolation *Salmonella* DNA from plants was done using kit. The sequencing of invA isolated from plant samples also was done.

The qPCR was done for DNA samples isolated from wheat root, shoot and substrate liquid inoculated with *Salmonella* strains. The Fluorescence *In Situ* Hybridization was done for inoculated plant samples. It was used specific probes for detection *Salmonella* by CLSM.

The results show that both investigated *Salmonella* strains were able to colonize wheat plants. The number of *Salmonella* DNA copies was 4.01×10^6 per 1 g root (*S. enteric*a) and 3.32×10^7 per 1 g root (*S. typhimurium*).

1 Introduction

In recent years, there is an increasing number of outbreaks caused by consumption fresh vegetables which are contaminated with human pathogen bacteria. The application organic fertilizers and contaminated irrigation water are the main reasons of contamination by pathogen bacteria during the food production.

The consumption of vegetables is very important for prevention cardiovascular diseases and it is recommended by World Health Organization. The fresh vegetables are essential for healthy nutrition and provide minerals and vitamins. The vegetables are mostly consumed raw and it is very important to avoid its microbiological contamination during the production chain.

Recently there has been an increasing number of outbreaks caused by contaminated fresh vegetables. According to CDC (Center of Disease Control and Prevention, 2010), the lettuce was the one of the most frequent source of foodborne outbreaks in USA during the 2007.

Today, there are many methods for detecting human pathogen bacteria in food. The conventional methods which are based on cultures are generally timeconsuming and new methods are needed to exceed their performance.

Immunology-based methods for detection human pathogen are very powerful tools and they provide extraction pathogen from bacterial suspension usingantibody coated magnetic beads. The PCR methods give more conclusive results, especially recent advances in PCR technology.Thus, with Real-Time-PCR (RT PCR), we are able to get very precise results only in few hours. Today, the most common methods for human pathogen detection in fresh vegetables are: colony counting technique, PCR and immunology-based methods. The PCR is much faster than other techniques and it takes approximately 6 to 24 hours to get result and this method does not need and include any previous enrichment steps. On the other hand, with RT PCR we can get results faster, only in few hours.

The general (conventional) PCR protocol for detection human pathogen bacteria in samples includes: denaturation of DNA, annealing of sequence specific primers, extension by polymerase 25–40 cycles. The PCR product can be analysed by gel electrophoresis or DNA sequencing. The qPCR is technique for amplification and simultaneously quantification a targeted DNA molecule.The qPCR allows detection and quantification of DNA sequence in real time after amplification cycle. The quantification includes fluorescent dyes which insert with double-stranded DNA during PCR oligonucleotide probes that illuminate after hybridization with complementary DNA and extension. The qPCR is combination of the amplification DNA and quantification of amplified DNA in real time. Also, it is possible to use probes which are labeled with different dyes and they allow quantification and detection of multiple target genes in one PCR reaction.

In comparing with conventional microbiology methods, the PCR technique is much faster and requires less time to achieve precise and valid results. The advantage of PCR is detection of bacteria which are not able to grown in culture. The PCR allows increasing speed, sensitivity, specificity of detection human pathogen bacteria in fresh and ready to eat vegetables.

Today, it is known two qPCR methods, TaqMan and SYBR Green. The TaqMan PCR is based on fluorescent probes which must be selected according to very strict conditions and it cannot be always applied. The SYBR Green qPCR provides fast result compared to others technique and detection is based on binding of SYBR-Green dye into double stranded PCR products. It could be applied without the needing for probes linked to fluorescent molecules.

This study tries to develop protocol for rapid and precise detection of human pathogen bacteria in contaminated plants using the qPCR. The aim of this work is to develop and improve microbiological laboratory analysis of human pathogens using real-time PCR, develop a PCR method and develop a validation of protocol for it. It is very important to establish simple and reliable qPCR method which is using SYBR Green that could be suitable for routine analyses of *Salmonella spp.* in plants and fresh vegetables.

In general, the aim is application of the real time qPCR method for detection of*Salmonella enteric*a subsp. Welteweden and *Salmonella typhimurium* LT2 in wheat seedlings and to get an expertise with pathogen detection methods using real time qPCR techniques.

2 Materials, Methods and Notes

Bacterial strains which are used were *Salmonella enterica* subsp. Welteweden and *Salmonella typhimurium* LT2. The model plant for inoculation was wheat. The sterile wheat seed was incubated on NB plates at 30° C for 3 days, letting them germinate and plants were grown on quartz sand in sterile conditions. The bacterial suspension which is applied for inoculation seeds was ≈ 10^8 CFU (OD_{600} = 0.7) for both *Salmonella* strains. Before inoculation, the seedlings were washed in sterile H_2O five (5) times and they were kept in bacterial (*Salmonella*) solution 1 hour at 20°C before planting. The inoculated plants left to grow in phytochamber for 3 weeks.

The standard PCR was done for pure culture of *Salmonella enterica* subsp. Welteweden and *Salmonella typhimurium* LT2. The standard PCR included extraction bacterial (*Salmonella*) DNA using the Genomic DNA From Tissue kit, NucleoSpin Tussue (Machery-Nagel, www.mn-net.com). It was used specific primers for *Salmonella*: rfbJ; fliC; fijB; invA and hilA (for *Salmonella typhimurium* LT2). It was used 16 S PCR Program: Hotstart at 94°C – 5 min.; Denaturating at 94°C – 45 sec.; Annealing at 54°C – 45 sec.; Elongation at 72°C – 45 sec.; Final elongation at 72°C – 5 min.; Store at 4°C – continous (30 cycles). The electrophoresis was done in 1% Agarose in TAE Buffer + 3µl EtBr/100 ml at 120V; 400mA; 100W; 45 minutes.

Also, FISH (Fluorescence *In Situ* Hybridization) was done for pure cultures of *Salmonella enterica* subsp. Welteweden and *Salmonella typhimurium* LT2. The cultures of *Salmonella* strains incubated overnight in NB Broth at 37°C with shaking. After that, it was done fixation in Paraformaldehyde (PFA) (for G^- bacteria) according to Protocol Fixation Of Bacterial Liquid Cultures. The

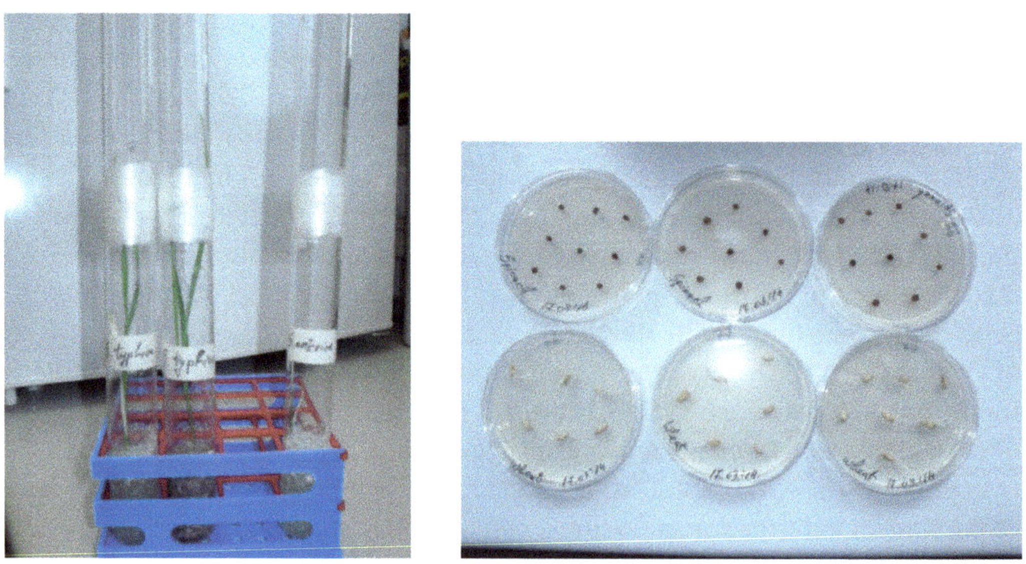

Figure 1: The wheat plant in glass tubes and seed germination in Petri dish.

Primer	**Target gene**	**Primer length (bp)**	**Sequence**	**Size of amplified fragment (bp)**	***Numbers designate Genbank-EMBL-DDBL ID numbers of the sequences in databases***
RfbJ-s RfbJ-as	*rfbJ*	24 24	5'-CCAGCACCAGTTCCAACTTGATAC-3' 5'-GGCTTCCGGCTTTATTGTTAAGCA-3'	663	AE008792
FliC-s FliC-as	fliC	24 24	5'-ATAGCCATCTTTACCAGTTCCCCC-3' 5'-GCTGCAACTGTTACAGGATATGCC-3'	183	D13689
FljB-s FljB-as	fljB	24 24	5'-ACGAATGGTACGGTCTCTGTAACC-3' 5'-TACCGTCGATAGTAACGACTTCGG-3'	526	AF045151
139-s 141-as	*invA*	26 22	5'-GTGAAATTATCGCCACGTTCGGGCAA-3' 5'-TCATCGCACCGTCAAAGGAACC-3'	284	Malorny at al. (2003)

Table 1: Primers for *Salmonella strains.*

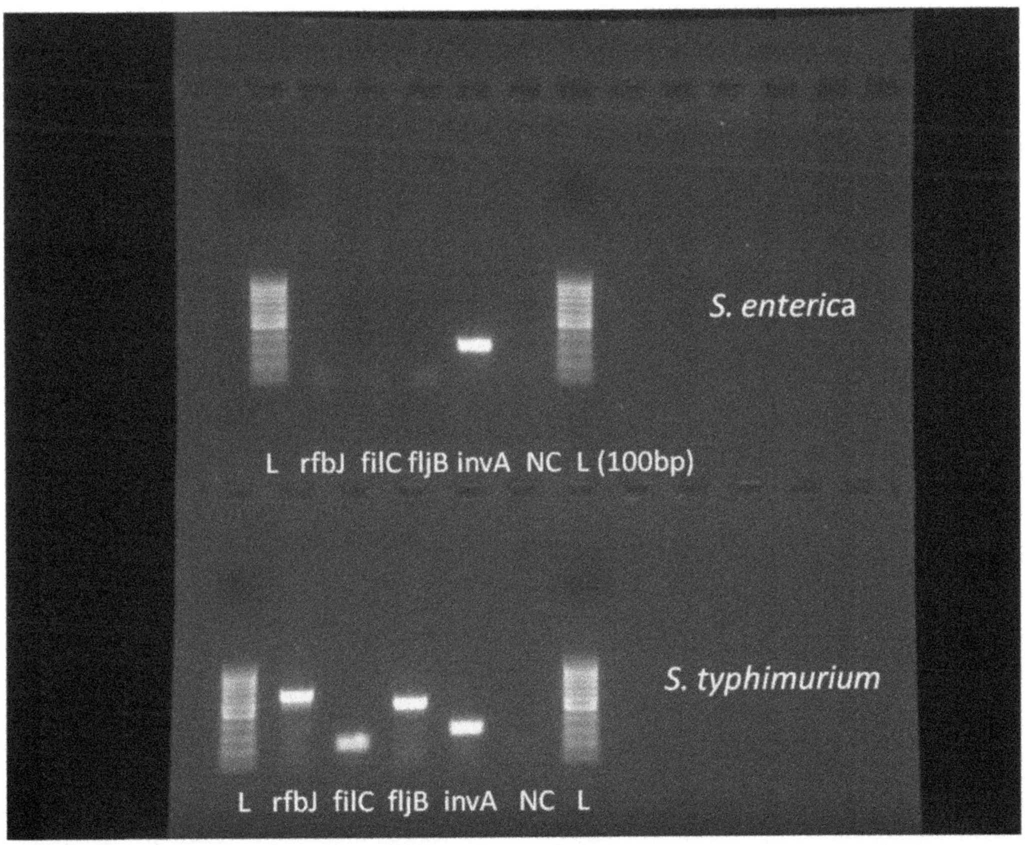

Figure 2: Standard PCR for pure culture of *Salmonella*.

oligonucleotide probes used in this analyze were: Salm-63-Cy3; Gam42a-Fluos; Bet42a-Oligo. The FISH protocol is: add 1–10μl PFA-fixed sample onto glass slide; drying at 46°C; EtOH-dehydration in 50%, 80%, 100% for 3 min. each; air drying; add 8μl hybridization buffer and 1μl probe. The samples were observed by CLSM (confocal laser scanning microscope).

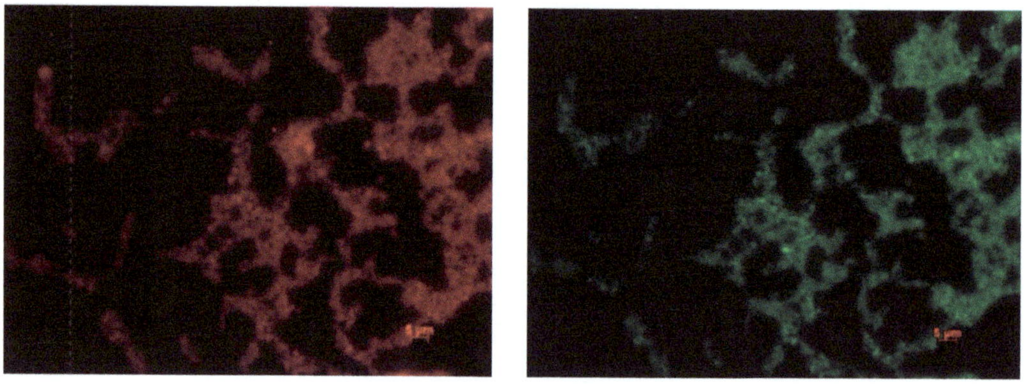

Figure 3: FISH for pure culture of *Salmonella typhimurium* LT2.

qPCR Analyses: In the goal to preparing standards for qPCR, cloning was done to get plasmid of *Salmonella* strains. PCR cloning was done for invA gene at *Salmonella* strains (*Salmonella enterica* subsp. Welteweden and *Salmonella typhimurium* LT2) according to PCR cloning protocol StrataClone PCR Clonong Kit (Stratagene). The protocol includes: Isolation plasmid with invA from *E. coli* competent cells (it is done by kit Plasmid DNA purification according to protocol NucleoSpin Plasmid QuickPure protocol); preparing standards for qPCR (it is calculated a number of moleculs per 1 μl of invA copies). After that, qPCR for DNA samples of *Salmonella* pure cultures was done according to 16 S PCR program and qPCR-samples were run on the gel.

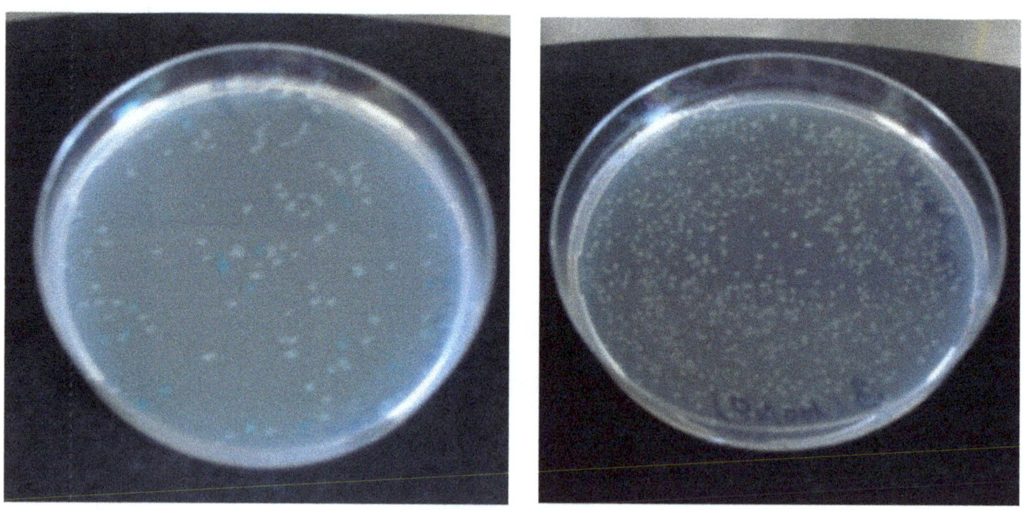

Figure 4: Colonies of transformed competent cells with plasmid of invA.

Isolation of Salmonella DNA from plants: The *Salmonella* DNA was isolated from:wheat root; wheat shoot (stem and leaves) (quartz sand) liquid. The protocol for isolation *Salmonella* DNA from wheat plants and substrate liquid includes: it is taken 0.5 g of plant material and 500 μl of liquid for analyses; the plant root and shoot were crashed with mortar and pestle in liquid nitrogen; after that, DNA from plant material and substrate liquid are further isolated according to FastDNA SPIN Kit for Soil (www.mpbio.com). Standard PCR for DNA isolated from plant and liquid samples was done to check presence of *Salmonella* DNA. The positive control is DNA isolated from *Salmonella* pure culture and negative control is reaction mixture without DNA.

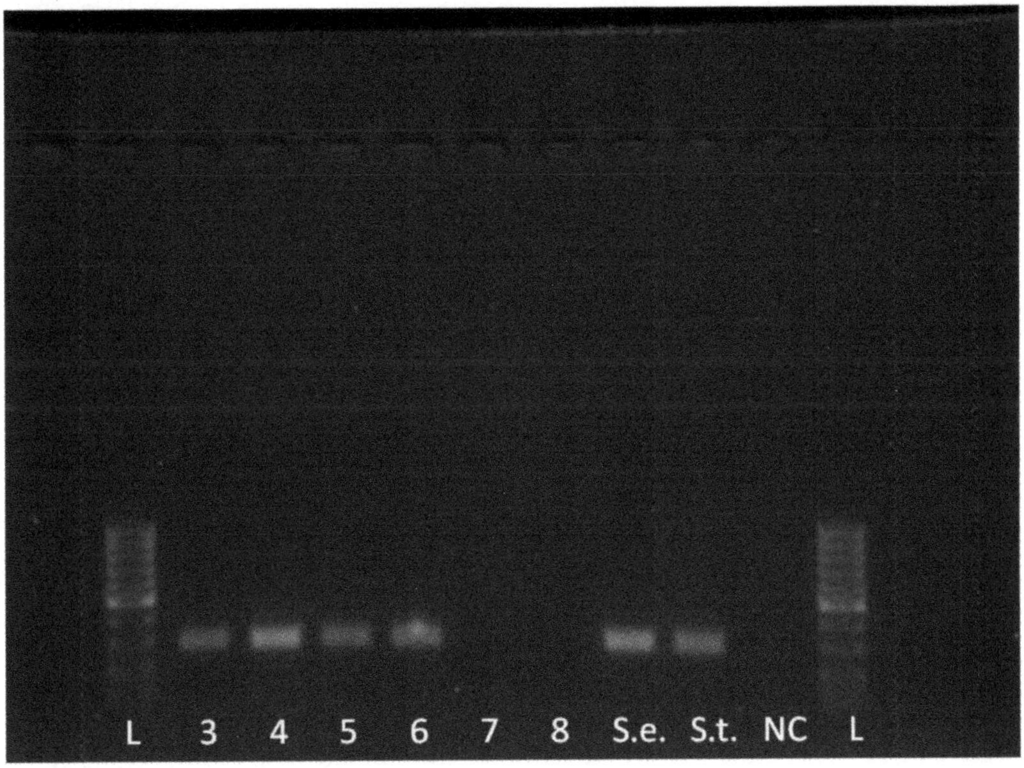

Figure 5: The standard PCR for DNA isolated from plant samples to check presence of *Salmonella* DNA (invA).

Legend: 3–wheat root inoculated by *Salmonella typhimurium* LT2; 4–wheat shoot inoculated by *S. typhimurium* LT2; 5–wheat root inoculated by *Salmonellaenterica* subsp. Welteweden; 6–wheat shoot inoculated by *S.enterica* subsp. Welteweden; 7–substrate liquid inoculated by *S. typhimurium* LT2; 8–substrate liquid inoculated by *S.enterica* subsp. Welteweden; S.e. and S.t.-positive control; NC-negative control.

Also, sequencing of invA isolated from plant samples was done and this analysis consisted of thefollowing: standard PCR; run gel; purification PCR mastermix samples; determination of DNA concentration by NanoDrop; Seq-PCR;

purification Seq-PCR product; putting samples in microtiter plate and doing sequencing.

Finally, the quantitative PCR (qPCR) was done for DNA samples isolated from: wheat root inoculated with *Salmonella typhimurium* LT2; wheat shoot inoculated with *Salmonella typhimurium* LT2; wheat root inoculated with *Salmonella enteric*a subsp. Welteweden; wheat shoot inoculated with *Salmonella enteric*a subsp. Welteweden; liquid inoculated with *Salmonella typhimurium* LT2 and liquid inoculated with *Salmonella enteric*a subsp. Welteweden. The DNA samples were run on the gel.

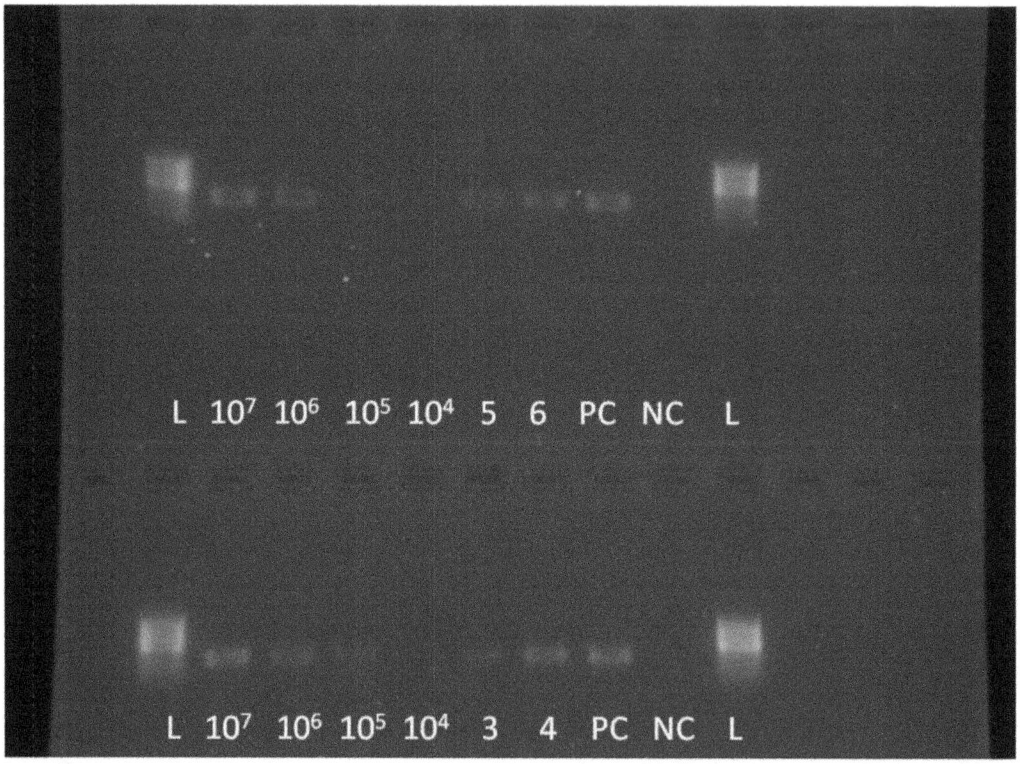

Figure 6: qPCR products of wheat samples inoculated with *Salmonella* strains (3; 4; 5; 6) on the gel.

Legend: 3–wheat root inoculated by *Salmonella typhimurium* LT2; 4–wheat shoot inoculated by *S. typhimurium* LT2; 5–wheat root inoculated by *Salmonellaenteric*a subsp. Welteweden; 6–wheat shoot inoculated by *S.enteric*a subsp. Welteweden; PC-positive control; NC-negative control.

Also, FISH was done for inoculated plant samples and CLSM analyses and sampleswere: wheat root, stem and leaf. The specific probes which were used for detection *Salmonella* strains by CLSM were: Salm 63 – Cy3 (red); Gam 42 – Fluos (green) and Bet 42 a – Oligo. The FISH analyses were done according to protocol: In Situ Hybridization Protocol for plant material.

The investigated *Salmonella* strains were able to colonize wheat plants. The number of *Salmonella* DNA copies was 4.01×10^6 per 1 g root (*S. enterica*) and 3.32×10^7 per 1 g root (*S. typhimurium*).

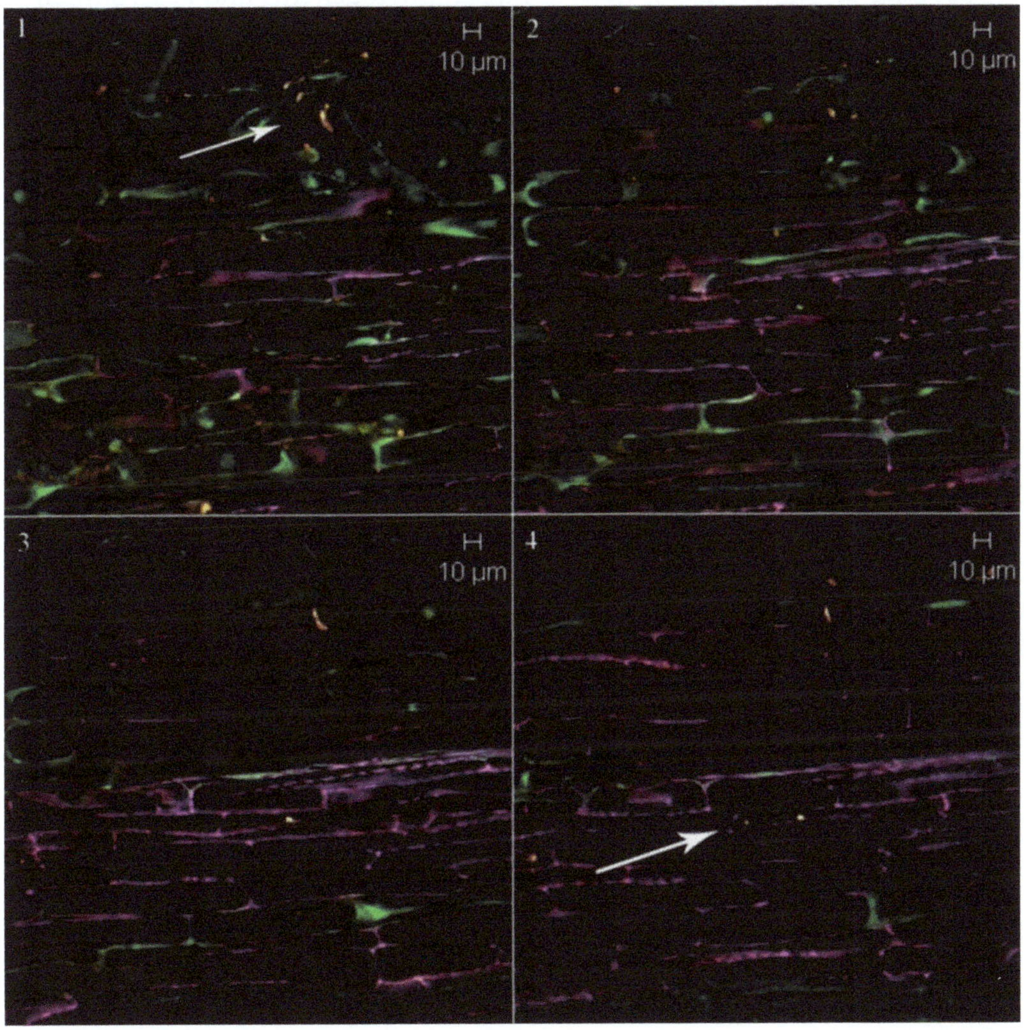

Figure 7: CLSM micrograph of endophyt colonization root by *Salmonella typhimurium*.

3 Acknowledgement

This study was supported by the EU Commission (FP7 project AREA) and Serbian Ministry of Education, Science and Technological Development (project TR 31080).

I am very thankful to Prof. Anton Hartmann, Dr Michael Schmid, Angelo Weiß and all people from Helmholtz Zentrum Munchen, Germany, AMP Research Unit Microbe-Plant Interactions who helped me to realize my research.

4 References

Amann, R. I., Krumholz, L. & Stahl, D. A. (1990). Fluorescent-oligonucleotide probing of whole cells for determinative, phylogenetic, and environmental studies in microbiology. *Journal of Bacteriology 172*, 762–770. DOI: https://doi.org/10.1128/jb.172.2.762-770.1990

Amann, R.I., Zarda, B., Stahl, D. A. & Schleifer, K. H. (1992). Identification of individual prokaryotic cells by using enzyme labeled, rRNA-targeted oligonucleotide probes. *Apply Environtal Microbiology 58*, 3007–3011.

Barak D. J., Whitehead L. C. & Charkowski A. O. (2003). Difference in Attachment of *Salmonella enterica* Serovars and *Escerichia coli* O157:H7 to Alfalfa Sprouts. *Apply Environtal Microbiology* 69 (8),4556–4560. DOI: https://doi.org/10.1128/AEM.68.10.4758-4763.2002

Beuchat, L. R. (2002). Ecological factors influencing survival and growth of human pathogens on raw fruits and vegetables. *Microbes and Infections 4*, 413–423. DOI: https://doi.org/10.1016/S1286-4579(02)01555-1

Beuchat, L. R., & Ryu, J. H. (1997). Produce handling and processing practices. *Emerging Infectious Diseases. 3*,459–465. DOI: https://doi.org/10.3201/eid0304.970407

Brandl, M. T. & Mandrell. R. E. (2002). Fitness *of Salmonella enterica* serovar Thompson in the cilantro phyllosphere. *Apply Environtal Microbiology*. 68, 3614–3621. DOI: https://doi.org/10.1128/AEM.68.7.3614-3621.2002

Burun B. & Coban Poyrazoglu, E. (2002). Embryo Culture in Barley (*Hordeum vulgare* L.). *Turkish Journal of Biology 26*, 175–180.

CDC (2006). Ongoing multistate outbreak of Escherichia coli serotype O157:H7 infections associated with consumption of fresh spinach – United States, September 2006. Morb Mortal Wkly Rep 55, 1045–1046. https://www.cdc.gov/mmwr/preview/mmwrhtml/mm55d926a1.htm

Charkowski, A. Barak, O.J.D., Sarreal, C. Z. & Mandrell, R. E. (2002). Differences in growth of *Salmonella enterica* and *Escherichia coli* O157:H7 on alfalfa sprouts. *Apply Environtal Microbiology 68(6),* 3114–3120. DOI: https://doi.org/10.1128/AEM.68.6.3114-3120.2002

Cooley, M. B., Miller, W. G. & Mandrell, R. E. (2003). Colonization of *Arabidopsis thaliana* with *Salmonella enterica* or enterohemorrhagic *Escherichia coli* O157:H7 andcompetition by *Enterobacter asburiae. Apply Environtal Microbiology 69*, 4915–4926. DOI: https://doi.org/10.1128/AEM.69.8.4915-4926.2003.

Cummings, K., Barrett, E., Mohle-Boetani, J. C., Brooks, J. T., Farrar, J.T., Hunt, Fiore, A., Komatsu, K., Werner, S. B. & Slutsker, L. (2001). A multistate outbreak of *Salmonella enterica* serotype baildon associated with domestic raw tomatoes. *Emerging Infectious Diseases 7*, 1046–1048. DOI: https://doi.org/10.3201/eid0706.010625

Dong, Y, Iniguez, A. L. Ahmer, B. M. & Triplett, E. W. (2003). Kinetics and strain specificity of rhizosphere and endophytic colonization by enteric bacteria on seedlings of *Medicago sativa* and *Medicago truncatula. Apply Environtal Microbiology 69*, 1783–1790, DOI: https://doi.org/10.1128/AEM.69.3.1783-1790.2003

Doyle, M. P., & Schoeni. J. L. (1986). Isolation of *Campylobacter jejuni* from retail mushrooms. *Apply Environtal Microbiology 51*, 449– 450.

Eldor, P. (2007). *Soil Microbiology, Ecology and Biochemistry.* Oxford: Academic Press is an imprint of Elsevier, ISBN: 9780123914118

Gandhi, M., Golding, S. Yaron, S. & Matthews, K. R. (2001). Use of green fluorescent protein expressing *Salmonella* Stanley to investigate survival, spatial location, and control on alfalfa sprouts, *Journal of food protection 64*, 1891–1898. DOI: https://doi.org/10.4315/0362-028X-64.12.1891

Gillespie, I.A. (2004). Outbreak of Salmonella Newport infection associated with lettuce in the UK. Eurosurveillance Weekly 8, http://www.eurosurveillance.org/ViewArticle.aspx?ArticleId=2562

Guo, X. Chen, J., Brackett, R. E. & Beuchat, L. R. (2001). Survival of salmonellae on and in tomato plants from the time of inoculation at flowering and early stages of fruit development through fruit ripening. *Apply Environtal Microbiology 67*, 4760–4764. DOI: https://doi.org/10.1128/AEM.67.10.4760-4764.2001

Guo, X., van Iersel, M. W., Chen, J., Brackett, R. E. & Beuchat, L. R. (2002). Evidence of association of salmonellae with tomato plants grown hydroponically in inoculated nutrient solution. *Apply Environtal Microbiology 68* (7), 3639–3643. DOI: https://doi.org/10.1128/AEM.68.7.3639-3643.2002

Hedberg, C. W., Angulo, F. J., White, K. E., Langkop, C. W., Schell, W. L., Stobierski, M. G., Schuchat, A., Besser, J. M., Dietrich, S., Helsel, L., Griffin, P. M., McFarland, J. W. & Osterholm, M. T. (1999). Outbreak of salmonellosis associated with eating uncooked tomatoes: implications for public health. *Epidemiology & Infection 122*, 385–393. DOI: https://doi.org/10.1017/S0950268899002393

Horby, P.W., O'Brien, S.J., Adak, G.K., Graham, C., Hawker, J.I., Hunter, P., Lane, C., Lawson, A.J., Mitchell, R.T., Reacher, M.H., Threlfall, E.J. & Ward, L.R. (2003). A national outbreak of multiresistant Salmonella enterica serovar Typhimurium definitive phage type (DT) 104 associated with consumption of lettuce. *Epidemiology & Infection 130*, 169–178, DOI: https://doi.org/10.1017/S0950268802008063

Islam, M., Morgan, J., Doyle, M.P., Phatak, S.C., Millner, P. & Jiang, X. (2004). Fate of *Salmonella enterica* serovars Typhimurium on carrots and radishes grown in fields treated with contaminated manure composts or irrigation water. *Apply Environtal Microbiology 70*, 2497–2502. DOI: https://doi.org/10.1128/AEM.70.4.2497-2502.2004

Jablasone, J., Warriner, K. & Griffiths, M. (2005). Interactions of *Escherichia coli* O157:H7, *Salmonella Typhimurium* and *Listeria monocytogenes* in plants cultivated in a gnotobiotic system. *Internatiol Journal of Food Microbiology 99*, 7–18.

Jerngklinchan, J. & Saitanu, K. (1993). The occurrence of salmonellae in bean sprouts in Thailand. *The Southeast Asian Journal of Tropical Medicine and Public Health 24*, 114–118.

Manz, W., Amann, R., Ludwig, W., Wagner, M. & Schleifer, K.H. (1992). Phylogenetic oligodeoxynucleotide probes for the major subclasses of Proteobacteria: problems and solutions. *Systematic and Applied Microbiology 15*, 593–600. DOI: https://doi.org/10.1016/S0723-2020(11)80121-9

Moter, A., & Gobel, U. (2000). Fluorescence in situ hibridization (FISH) for direct visualisation of microorganisms. *Journal of Microbiological Methods 41(2)*, 85–112. DOI: https://doi.org/10.1016/S0167-7012(00)00152-4

Natvig, E.E., Ingham, S.C., Ingham, B.H., Cooperband, L.R. & Roper, T.R. (2002). *Salmonella enterica* serovar Typhimurium and *Escherichia coli* contamination of root and leaf vegetables grown in soils with incorporated bovine manure. *Apply Environtal Microbiology 68*, 2737–2744. DOI: https://doi.org/10.1128/AEM.68.6.2737-2744.2002

Pezzoli, L., Elson, R., Little, C., Fisher, I., Yip, H., Peters, T., Hampton, M., De Pinna, E., Coia, J.E., Mather, H.A., Brown, D.J., Nielsen, E.M., Ethelberg, S., Heck, M., de Jager, C. & Threlfall, J. (2007). International outbreak of Salmonella Senftenberg in 2007. Eurosurveillance Weekly 12, (accessed on 15/06/07) http://www.eurosurveillance.org/ViewArticle.aspx?ArticleId=3218

Raina M. Maier, Ian L. Pepper, Charles P. Gerba., Blomme, B., & Handler, A. (2009). Enviromental Microbiology 2nd ed. Amsterdam; Boston: Elsevier/Academic Press, c2009 9780123705198

Smit, G., Kijne, J. W. & Lugtenberg, B. J. (1987). Involvement of both cellulose fibrils and a Ca2+-dependent adhesin in the attachment of Rhizobium leguminosarum to pea root hair tips. *Journal of Bacteriology 169(9)*, 4294–4301. DOI: https://doi.org/10.1128/jb.169.9.4294-4301.1987

Soderstrom, A., Lindberg, A. & Andersson, Y. (2005). EHEC O157 outbreak in Sweden from locally produced lettuce, August–September 2005. *Eurosurveillance, 10 (9):* E050922.1 Retrieved from (http://www.eurosurveillance.org/ViewArticle.aspx?ArticleId=2794).

Takkinen, J., Nakari, U-M., Johansson, T., Niskanen, T., Siitonen, A. & Kuusi, M. (2005). A nationwide outbreak of multiresistant *Salmonella Typhimurium* var. Copenhagen DT104B infection in Finland due to contaminated lettuce from Spain, May 2005. *Eurosurveillance Weekly 10(26)*. Retrieved from http://www.eurosurveillance.org/ViewArticle.aspx?ArticleId=2734

Viswanathan, P. & Kaur, R. (2001). Prevalence and growth of pathogens on salad vegetables, fruit and sprouts. *International Journal of Hygiene and Environtal Health 203*, 205–213. DOI: https://doi.org/10.1078/S1438-4639(04)70030-9

Warriner, K., Spahiolas, S., Dickinson, M., Wright, C. & Waites. W. M. (2003). Internalization of bioluminescent *Escherichia coli* and *Salmonella* Montevideo in growing bean sprouts. *Journal Apply Microbiology 95*, 719–727. DOI: https://doi.org/10.1046/j.1365-2672.2003.02037.x

Zogaj, X., Nimtz, M., Rohde, M., Bokranz, W. & Romling. U. (2001). The multicellular morphotypes of *Salmonella Typhimurium* and *Escherichia coli* produce cellulose as the second component of the extracellular matrix. *Molecular Microbiology 39*, 1452–1463. DOI: https://doi.org/10.1046/j.1365-2958.2001.02337.x

The Application of Molecular Methods in Diagnostics of Phytopathogenic Viruses, Fungi and Fungus–Like Organisms

Ivana Stanković and Ana Vučurović

Abstract

The aim of this chapter is to describe some of the protocols which are used in the diagnostics of plant viruses, fungi and fungus-like organisms. Special adaptation of certain protocols is required for each of this group of pathogens and even more, for every type of virus, fungus or fungus-like organism, each type of raw material in terms of plant organs, plant host species, time of year and conditions in the laboratory. The protocols of molecular methods for detection, identification and characterization of phytopathogenic organisms cited here in include detailed instructions for its application. This chapter contains specific recommendations for selection of plant material, extraction of total RNAs or DNAs, detailed procedure for RT-PCR or PCR depending on phytopathogenic organisms and instruction for interpretation of obtained results. The chapter also focuses on application of DNA cloning in plant virology. The text on DNA cloning contains instruction for basic steps including amplification of DNA fragment that we want to clone, ligation of PCR product with vector DNA, insertation of recombinant DNA into bacteria cells, the reproduction

How to cite this book chapter:
Stanković, I. and Vučurović, A. 2019. The Application of Molecular Methods in Diagnostics of Phytopathogenic Viruses, Fungi and Fungus–Like Organisms. In: Vucelić Radović, B., Lazić, D. and Nikšić, M. (eds.) *Application of Molecular Methods and Raman Microscopy/Spectroscopy in Agricultural Sciences and Food Technology*, Pp. 59–82. London: Ubiquity Press. DOI: https://doi.org/10.5334/bbj.e.

of bacteria together with the recombinant DNA, screening clones with recombinant DNA, and plasmid extraction and purification from the bacterial cells.

1 Molecular methods in plant virology and mycology

Identification of plant pathogens, in addition to conventional methods, requires the use of a variety of different molecular methods, which contribute to the accuracy, reliability, speed and efficiency in phytopathology at all. In addition, the risk of introduction of invasive plant pathogens whether cultivated, ornamental plants or in natural communities (forests, pastures) grows as a result of globalization, increased mobility of people, climate changes and pathogens and vector evolution (Anderon et al. 2004; Miller et al. 2009).

Generally speaking plant pathogens, whether they are expansive (emerging), re–emerging (e.g., new races, pathotypes, forms resistant to pesticides or antibiotics), and chronic/endemic pathogens that are known and present for long time, but still can prompt epidemics, can cause economically significant yield losses (Strange & Scott 2005). Because of this, the application of molecular methods is of great importance and it is impossible to get totally reliable results in the detection, identification and characterization of plant viruses, fungi and fungus like–organisms without the application of this group of methods (Miller et al. 2009).

Adequate and timely diagnosis of diseases and detection of pathogens are crucial for prompt protection of cultivated crops, as well as natural biocoenosis (forest ecosystems and pastures), or in order to undertake preventive or application of a small number of therapeutic measures. Errors during detection of pathogens and disease diagnosis may lead to applications of inadequate control measures, and thus to the reduction in yield or market value of the crop. Inadequate phytosanitary measures based on an irregular and out of time identification of causal agent of disease may lead to the introduction of new, invasive and quarantine pathogen if the pathogen is detected late, which can cause economically very negative impact on production in the country, as well as on the export of plant production products (Miller et al. 2009).

Conventional methods which are used in the detection of plant pathogens, are sometimes time consuming, tedious, and often require extensive experience and expertise of the person who is applies them. In the certain cases of closely related pathogens, conventional methods can lead to the wrong conclusion or to inaccurate levels of identification. Because of these limitations, several methods have been developed based on the properties of nucleic acids that make up the genome of plant pathogens. These methods are usually collectively referred to as molecular methods. There is a large number of molecular methods, and common to all of them is that they are fast (often significantly faster than conventional), reliable and specific. Usage of molecular methods allows the detection of pathogens in different parts of the host plants, as well as in different natural environments, such as water for drinking or irrigation or in soil (McPherson & Møller 2000; Reece 2004).

Molecular methods are used in the science for various purposes since the late 50's and early 60's of the 20th century, however the greates progress and the fastest development was achieved by Kary Mullis in 1983 with the invention of the *Polymerase Chain Reaction (PCR)*. This breakthrough changed molecular biology completlly, as well as virtually all biological and other sciences that rely on it, and deal with the research of genes and genomes. Since the invention of PCR, a number of modifications are made to particular usages, so today this method has a very wide range of use in all spheres of life, from medicine, through criminology, to the production of food and control of food safety (McPherson & Møller 2000).

PCR method has several basic components that are used in everyday laboratory work to create a large number of copies of a specific portion of DNA in laboratory tube. PCR actually functions as a DNA copier. Although seemingly simple, PCR is actually a complicated process which involves a large number of components/reactants. Some of them, such as the DNA matrix, at the beginning are present in a very low concentrations, but as the reaction is going on their concentration increases dramatically, while the concentration of some components (dNTP, primers) do not significantly changes during the process. Also, rapid changes in the temperature and the pH value have significant influence on the interaction of the molecules during the process of PCR. All this makes PCR at the same time very complicated, but it opens a large number of various possibilities for manipulation and analysis of DNA (McPherson & Møller 2000).

In the following text we have listed some of the protocols which are daily used in the research of plant viruses, fungi and fungus–like organisms, including detailed instructions, as well as the positive experience acquired during the research. Each of this group of pathogens, and even more every type of virus, fungus or fungus–like organism, each type of raw material in terms of plant organs, plant host species, time of year and the conditions in the laboratory requires special adaptation of certain protocols. The following text contains specific recommendations gained on the basis of experience in working with molecular methods during the research at the Laboratory of Virology and Mycology, Department of Plant Pathology, Faculty of Agriculture, University of Belgrade.

2 The application of molecular methods in detection of phytopathogenic viruses

2.1 Selection of plant material

When selecting a starting plant material for preparation of samples for molecular analysis possible uneven distribution of viruses in the plant must be taken into account. In order to improve accuracy of test and obtaining valid results, it is recommended to prepare compound samples of the smaller pieces of leaves with more expressed symptoms from many different parts of plant. In the case of necrotic symptoms, green tissue that borders necrotic parts should be used. Young, fresh and fully developed leaves are recommended for usage, older parts of plants are to be avoided. With the aging or necrosis of plant tissue virus concentration decreases and successful detection is harder (Krstić et al. 2008, 2010).

2.2 Extraction of Total RNA

Before reverse transcription followed by polymerase chain reaction (RT–PCR) total ribonucleic acid (RNA) should be extracted from infected plant material. For this purpose numerous protocols for total RNA extraction from plant material are developed. Like commercial kits: RNeasy Plant Mini Kit (Qiagen, Hilden, Germany; https://www.qiagen.com) and RNAqueous Small Scale Phenol–Free Total RNA Isolation Kit (Ambion, Inc., Applied Biosystems, USA; http://www.appliedbiosystems.com/http://www.thermofisher.com) which includes usage of columns with membranes on which nucleic acid is attaching. However, if extraction of total RNA is from seeds or plant species with high concentration of polyphenols or starch, which easily can clog membranes (McKirdy et al. 1998), CTAB (hexadecyltrimethylammoniumbromide) is recommended method (Bekesiova et al. 1999; Zeng and Yang 2002; Iandolino et al. 2004). Application of these and other different commercial kits is somewhat more expensive, but it requires less training and experience and usually gives uniformed results, while the use of CTAB method ensures that the extraction is better adopted to the specific characteristics of the starting material (Krstić et al. 2008, 2010).

2.2.1 Total RNA extraction using RNeasy Plant Mini Kit (Anonymous, 2012b)

Important Note:

1. The RNeasy Plant Mini Kit comprises two buffers: Buffer RLT and Buffer RLC, containing guanidine thiocyanate and guanidine hydrochloride, respectively. Buffer RLT is commonly used. In the cases when guanidine thiocyanate causes solidification of the sample Buffer RLC should be used.
2. If the Buffer RLT precipitates during storage, redissolve it by heating.
3. All steps of the procedure must be done quickly and at room temperature.
4. Add 10 μl of β–Mercaptoethanol per 1 ml of the buffer, stored buffer solution at room temperature (15–25°C). After addition of β–Mercaptoethanol, the buffer should be used for up to 1 month.

5. Before using Buffer RPE for the first time, add 4 ml ethyl alcohol (96–100%) per each ml of buffer.

Total RNA extraction protocol:

1. Choose 100 mg of the symptomatic plant material.
2. Homogenize chosen plant material in liquid nitrogen using, for example, a mortar and pestle. Transfer plant tissue together with liquid nitrogen into a 2 ml microcentrifuge tube and add 450 μl Buffer RLT or Buffer RLC, after liquid nitrogen was evaporated.
3. Vortex vigorously and incubate samples at 56°C for 1–3 min.
4. After incubation, transfer the sample to a QIAshredder spin column (lilac) in collection tube and centrifuge at full speed for 2 min. This centrifugation allows additional grind of plant material and retention of large parts of plant tissue.
5. Transfer only the supernatant in a new 2 ml microcentrifuge tube without disturbing pellet forms during centrifugation and add 225 μl of ethanol (96–100%). Mix by pipetting.
6. About 650 μl of the sample transfer to an RNeasy spin column (pink) in a 2 ml collection tube and centrifuge at ≥10,000 rpm for 15 s.
7. Discard the liquid from collection tube and add 700 μl Buffer RW1 to the RNeasy spin column.
8. Centrifuge sample at ≥10,000 rpm for 15 s and discard the fluid from collection tube.
9. In order to wash membrane, add 500 μl Buffer RPE to the RNeasy spin column and centrifuge at ≥10,000 rpm for 15 s. Discard the fluid from collection tube and reuse it in the next step.
10. Pipet new 500 μl Buffer RPE to the RNeasy spin column and centrifuge at ≥10,000 rpm for 2 min.
11. To remove residues of ethanol or extraction buffer, transfer the RNeasy spin column in a new 2 ml collection tube and centrifuge at full speed for 1 min.
12. In order to elute RNA, transfer the RNeasy spin column in a new 1.5 ml microcentrifuge tube and add 30–50 μl RNase–free water. Centrifuge at ≥10,000 rpm for 1 min.
13. Store isolated RNA at −20°C or −80°C.

2.2.2 Total RNA extraction using RNAqueous Small Scale Phenol–Free Total RNA Isolation Kit (Anonymous, 2008)

Important Note:

1. Prior to the first use, add 100 ml of 38.4% ethanol in water to obtain a 64% ethanol solution.

2. Wash Solution #2/3 Concentrate is supplied as a concentrate. Before using for the first time, add 64 ml 100% ethaol to obtain a working solution.
3. Briefly inspect the Filter Cartridges before use. If glass fiber filters are dislodged gently push the filter down to the bottom of the cartridge using RNase–free pipette tip.
4. Befor use heat Elution Solution in an RNase–free microcentrifuge tube in a heat block set to 70–80°C.
5. Perform all centrifugation steps at 20–25°C in a standard microcentrifuge. Ensure that the centrifuge does not cool below 20°C.

Total RNA extraction according follow protocol:

1. Weigh 60 mg of plant material with syptoms.
2. Place the weighed tissue in liquid nitrogen, and grind thoroughly with a mortar and pestle. Allow the liquid nitrogen to evaporate, but do not allow the tissue to thaw. Add 12 volumes (720 μl) Lysis/Binding Solution and 60 μl Plant RNA Isolation Aid in completely homogenized material.
3. Transfer sample in to 2 ml microcentrifuge tube and centrifuge for 2–3 min at 10000–14000 rpm.
4. Carefully transfer the supernatant in to new 2 ml microcentrifuge tube without disturbing the cell–debris pellet in the collection tube, and then add 720 μl 64% ethanol. Mix the sample using pipette.
5. Apply the ethanol mixture to a Filter Cartridge assembled in a Collection Tube (the maximum volume that can be applied at one time is approximately 700 μL). Close the lid gently, and centrifuge for 1 min at 1000–14000 rpm (13000 rpm). Discard the flow–through and reuse the Collection Tube.
6. Repeat this step until the entire sample has been drawn through the filter. **Note:** Maximum 2 mL of sample mixture can be passed through the filter without clogging or exceeding its RNA binding capacity.
7. Apply 700 μL Wash Solution #1 to the Filter Cartridge. Close the lid gently, and centrifuge for 1 min at 1000–14000 rpm (13000 rpm). Discard the flow–through and reuse the Collection Tube.
8. Apply 500 μL Wash Solution #2/3 to the Filter Cartridge. Close the lid gently, and centrifuge for 1 min at 1000–14000 rpm (13000 rpm). Discard the flow–through and reuse the Collection Tube.
9. Repeat this step using new 500 μL Wash Solution #2/3.
10. Discard the flow–through, centrifuge Filter Cartridge for 1 min at 1000–14000 rpm (13000 rpm).
11. Pipet 40 μl Elution Solution preheated to 70–80°C to the center of the Filter Cartridge and centrifuge for 30 s at 1000–14000 rpm (13000 rpm).
12. Repeat this step using new 10 μl Elution Solution.
13. Isolated RNA store at –80°C.

2.2.3 Total RNA extraction using CTAB method (Bekesiova et al. 1999)

Important note:

1. Using CTAB method homogenization is done with 2% CTAB, 2% PVP K 25, 100 mM Tris–HCL, 25 mM Na–EDTA and 2 M NaCl extraction buffer pH 8.0. Prepared buffer is stored at room temperature under diffuse light.
2. Before using extraction buffer, it should be added 20 µl of β–Mercaptoethanol (β–ME) per 1 ml of the buffer. After addition of β–ME, the buffer must be stored at room temperature (15–25°C) and should be used for up to 1 month.
3. Before isolation prepare chlorophorm:isoamyl alcohol mixture in 24:1 proportion, 10 M LiCl and 3 M sodium acetate pH 5.2.
4. Perform all centrifugation steps at 4°C.
5. Transfer certain amount of extraction buffer into microcentrifuge tube and preheat it at 65°C in a water bath.

Total RNA extraction protocol:

1. Use up to 100 mg of plant material.
2. Homogenize plant tissue in liquid nitrogen with a mortar and pestle. Transfer plant tissue together with liquid nitrogen into a 2 ml microcentrifuge tube and add 700 µl of preheated extraction buffer, after liquid nitrogen was evaporated.
3. Vortex vigorously. Incubation for 10 min at 56°C may help to disrupt the tissue, periodically shake microtube.
4. After incubation add 700 µl of chloroform: isoamyl alcohol mixture in 24:1 proportion. Centrifuge at 4°C for 10 min at 10 000 rpm.
5. After centrifuge three phases are phormed. Tranfer top phase, which contains RNA, into new 2 ml microtube and add another 700 µl of chloroform: isoamyl alcohol mixture in 24:1 proportion.
6. Centrifuge at 4°C for 10 min at 10 000 rpm.
7. After cetrifuge transfer top phase into new 2 ml microtube and add 175 µl 10 M LiCl. Incubate over night at 4°C.
8. Centrifuge at 4°C for 20 min at 10 000 rpm to form RNA pellet. Discard supernatant, pellet disolve by adding 50 µl of DEPC water.
9. Precipitate RNA by adding 70 µl of 3 M sodium acetate (pH 5.2), 1750 µl 96% ethanol and 30 min incubation at –70°C.
10. After incubation centrifuge at 4°C for 20 min at 10 000 rpm. Discard supernatant, wash pellet by adding 1 ml of 75% ethanol.
11. Centrifuge at 4°C for 5 min at 10 000 rpm to form RNA pellet. Discard supernatant, let pellet to dry at room temperature.

12. After all ethanol evaporate disolve RNA in 30 µl of RNase free water by mixing it with pipette.
13. Isolated RNA store at –80°C.

2.3 RT–PCR method

The vast majority of plant viruses are single–stranded RNA viruses with plus–sense polarity and for its detection, using PCR, are necessary to translate previously viral RNA into complementary DNA strand (cDNK). This step is referred to as reverse transcription or RT step.

RT step and PCR may be carried out as two separate processes in a single step or one after other in the same reaction tube. Application of „One–step" RT–PCR protocol presumes combination of two enzymes and has several advantages compared to protocol that takes place in two stages. The main advantage is that the reverse transcription of isolated RNA to cDNA and the amplification of cDNA itself taking place in the same tube, thus reducing the risks of possible contamination of the test sample. The continuous RT–PCR, in which RT and PCR methods make continuous reaction, is more sensitive from protocol that takes place in two separate steps (Sellner & Turbet 1998).

Procedure for OneStep RT–PCR Kit (Anonymous, 2012a)
Important notes before starting:

- The QIAGEN OneStep RT–PCR Enzyme Mix contains HotStarTaq DNA Polymerase that must be activated befor amplification. Incubation at 95°C for 15 min activates this polymerase but also inactivates the reverse transcriptases.
- Using The QIAGEN OneStep RT–PCR Kit with gene–specific primers at a final concentration of 0.6 µM, amplification of nonspecific products is avoided.
- All reactions must be carried out on ice and preheat thermal cycler to 50°C before put samples in it.
- Final concentration of $MgCl_2$ in the reaction mix is 2.5 mM, which provides adequate results.
- Preparation of reaction mix must be done in sterile, RNase–free area, separated from that used for RNA isolation or PCR product analysis.
- To avoid cross–contamination, tips with hydrophobic filters should be used.

Instructions and preparation
Preparation and storage of QIAGEN OneStep RT–PCR kit

1. The QIAGEN OneStep RT–PCR should be stored immediately upon receipt in laboratory at -20°C in a constant–temperature freezer.

2. The QIAGEN OneStep RT–PCR Enzyme Mix always stays in a freezer at –20°C. Do not centrifuge and vortex. As soon as you take the necessary amount for PCR–mix, return it into the freezer.
3. The 5x QIAGEN OneStep RT–PCR Buffer and dNTP Mix are divided into smaller portions that are stored in a freezer at –20°C. 5x QIAGEN OneStep RT–PCR Buffer divided into tubes per 50 μl, dNTP Mix per 10 μl. Quantities that are used daily keep in a refrigerator at 4°C.
4. RNase–free water when the first thaw out, kept in a refrigerator at 4°C.
5. Each ingredient of Kit (except enzyme mix), before putting PCR–mix, is briefly vortex (vortex is especialy necessary for $MgCl_2$ because it affect of its activity) and centrifuge (10 s at 5000 rpm/min). When all ingredients of PCR–mix is poured into the tube, it is briefly vortex or mixed by pipetting while adding each reagent before PCR–mix is pour into the tubes for individual samples.

Procedure for RT–PCR:

1. To avoid localized differences in concentration, thaw isolated RNA, primers, dNTP Mix, and 5x QIAGEN OneStep RT–PCR Buffer, vortex vigorously and centrifuge at 5000 rpm for 10 seconds. After that put all components on ice.
2. Prepare a reaction mixture using all the components required for RT–PCR, except the template RNA, which in its composition and quantity corresponds to the recommendations from Table 1. A negative control

Component	Volume/reaction	Final concentration
Master mix		
RNase–free water (provided)	Variable	–
5x QIAGEN OneStep RT–PCR Buffer	10 μl	1x
dNTP Mix (containing 10 mM of each dNTP)	2.0 μl	400 μM of each dNTP
Primer A	3.0 μl	0.6 μM[1]
Primer B	3.0 μl	0.6 μM[1]
QIAGEN OneStep RT–PCR Enzyme Mix	2 μl	–
Template RNA (added at step 4)	Variable	1 pg–2 μg/reaction
Total volume	50 μl	–

Table 1: Reaction components for one–step RT–PCR mix in volume of 50 μl.

[1]A final concentration of 0.6 μM of primers in the reaction mixture produces satisfactory results in most cases. However, sometimes using other primer concentrations (0.5–1.0 μM) may give better results.

(RNase free water) and a positive control (RNA of the reference isolate) should be included in each experiment.

Note: If it is not possible to measure the extracted RNA concentration using a spectrophotometer, it is suggested to add 2 μl RNA sample to 50 μl of reaction volume.

1. Mix the master mix thoroughly, and dispense appropriate volumes (48 μl) into PCR tubes, for each sample.
2. Add template RNA (≤ 2 μg/reaction) to the individual PCR tubes. Between samples must be altered extensions pipette and carefully handle the microtubes to avoid the formation of aerosols and possible contamination.
3. Put microtubes into the thermal cycler programmed according to the conditions shown in Table 2. The temperatures and cycling times, as well as number of cycling depend on virus–primers combination and the conditions are necessary to adjust for each specific primer pair.

Conditions in Thermal cycler	Duration	t°C	Number of cycles
Reverse transcription	30 min	50°C	
Initial PCR activation	15 min	95°C	
Denaturation	0.5–1 min	94°C	
Annealing	0.5–1 min	50–68°C	25–40
Extension	1 min	72°C	
Final extension	10 min	72°C	

Table 2: Thermal cycler conditions for application of QIAGEN OneStep RT–PCR Kit.

Note: Before amplification, hold the microtubes on the ice until Thermal cycler is warmed to 50°C.
Note: After amplification, store samples at 2–8°C overnight, or at –20°C for longer period.

3 The application of molecular methods in detection of phytogenetic fungi and fungus–like organisms

3.1 Selection of plant material for DNA extraction

Detection of phytogenetic fungi and pseudofungy by PCR method could be done after DNA extraction directly from plant tissue, leaf or branches, or from pure mycelium culture which are identified based on morphological characteristics.

In both cases DNA could be efficiently extracted by using described method. If DNA extraction is done from pure mycelium culture they need to be grown in liquid culture and segregate mycelia from agar in nutrient medium.

3.2 Growing mycelium for DNA extraction

There are a few different liquid mediums for growing mycelium which are usually adopted to the specific characteristics of fungi or pseudofungy which is subject of research. The most commonly used are potato dextrose broth (PDB) and pea broth (PB).

PDB (potato dextrose broth) is prepared of 200 g of potato and 20 g of dextrose in 1 l distillated water, all together autoclaved 15 min at 121°C. Diffuse 150 ml of PDB in Erlenmeyer bulb and sterilize, then sow with five fragments of colony (1 cm^2) from older cultures, reared on the PDB or other medium at 24°C in the dark. Seeded Erlenmeyer bulbs incubate for 15 days in the dark, at 24°C occasionally mixing with horizontal rotation. Liquid cultures of selected isolates are filtered over a layer of filter paper, then collect mycelium and dry under vacuum. Divide dried mycelia in parts of 100 mg, froze at –80°C, and keep in that condition until use (Konstantinova et al. 2002).

PB (pea broth) liquid medium is prepared of 120 g of frozen pea and 1 l of distillated water, all together autoclaved 15 min at 121°C. After sterilization filter medium and diffuse in prepared Erlenmeyer bulbs, 150 ml per isolate. Again sterilize Erlenmeyer bulbs by 15 min autoclaving at 121°C and seed fragments of pure 7 days old cultures, tested by rearing on CPA at 20°C in dark. Incubation of seeded Erlenmeyer bulbs is carried out at room temperature in terms of natural shifts day and night during 7 days. After the expiry of incubation collect developed mycelium without fragments of agar and keep at –80°C until extraction (Kroon et al. 2004).

3.3 DNA extraction

Method of polymerase chain reaction is preceded by extraction of total desoxyribonucleic acid (DNA). There are several protocols for extraction, as well as commercial kits. The most used are methods using DNeasy Plant Mini Kit (Qiagen, Hilden, Germany; https://www.qiagen.com) and standard CTAB (Cetyltrimethylamonium bromide) method (Day & Shattock 1997).

3.3.1 DNA extraction using DNeasy Plant Mini Kit (Anonymous, 2015a)

All components of DNeasy Plant kit, including RNase A liquid keep in dry condition, at room temperature (15–25°C), and under this condition they are stable for one year.

Note: This procedure is provided to process maximum of 100 mg of plant material. Using of larger quantities may adversely affect the success of extraction.

DNA isolation protocol with explanations:

Important notes before starting:

- Buffers AP1 and AP3/E concentrate may form precipitates and turn yellow upon storage. This does not affect its efficiency.
- All steps of centrifugation perform at room temperature (15–20°C) in a microcentrifuge.
- To redissolve Buffer AP3/E, warm it up to 65°C. Ethanol must be added after heating.
- To obtain working solutions of AW and AP3/E buffers, appropriate amount of ethanol (96–100%) should be added.
- Buffer AE should be warmed to 65°C.

1. Homogenize plant or fungal material using liquid nitrogen and mortar and pestle or tube and pestle. Transfer obtained homogenate to tube leaving liquid nitrogen to evaporate. Proceed immediately with procedure until the sample thawed.
2. Put 400 μl of AP1 Buffer and 4 μl of RNase A to 100 mg of ground tissue and mix it vigorously until no tissue clumps are visible. If needed, pipette supernatant to remove clumps because clumps are difficult to lyse and therefore could lower DNA yield.
3. To lyse the cells, put tube to incubation for 20 min at 65°C. During incubation, mix them 2–3 times manualy by flipping them up and down.
4. After incubation, add 130 μl of AP2 buffer and incubate mixture for 5 min on ice. In this step polysaccharides and proteins are being precipitated together with remain of detergent from previous steps.
5. In the case when lysate is very viscous, centrifuge it for 5 min at maximum speed. Pipette supernatant to QIAshredder spin column (lilac).
6. Centrifuge QIAshredder spin column for 2 min at maximum speed. After centrifugation, most precipitates and cell debris stays on column. Pipette liquid from collection tube making sure not to disturb pellet which could be formed.
7. Use a new tube to transfer approximately 450 μl of flow–through lysate from previous step.
8. Add Buffer AP3/E to lysate and mix it by pipetting, not vortex, immediately. For 450 μl lysate amoun of 675 μl of Buffer AP3/E must be added.
9. Transfer 650 μl of the mixture from previous step, toghether with precipitate could be formed, onto the new DNeasy mini spin column and centrifuge for 1 min at ≥6000 x g. Discard flow–through and use same collection tube in next step.
10. Repeat previous step using remaining sample.

11. Use a new 2 ml collection tube to place DNeasy column. Apply 500 μl of Buffer AW to the DNeasy column and centrifuge for 1 min at ≥6000 x g (≥8000 rpm). After centrifugation discard the flow–through and reuse the same collection tube in the next step.
12. Apply another 500 μl of Buffer AW to the DNeasy column and centrifuge for 2 min at maximum speed. After centrifugation discard collection tube together with flow–through.
13. Carefully transfer the DNeasy column to a 2 ml microcentrifuge tube (making sure that column does not contact with ethanol in collection tube) and pipette 100 μl of Buffer AE directly onto the DNeasy membrane (preheat Buffer AE to 65°C, previously). Leave tube for 5 min at room temperature. After incubation, centrifuge tube for 1 min at ≥6000 × g (≥8000 rpm) to elute DNA from membrane. If higher final concetration of DNA is needed, elute mambranes with 50 μl. If larger amounts of DNA are expected, elute mambranes with 200 μl.
14. If needed, previous step could be repeated in the same or in a new tube.
15. Isolated DNA is stored at −20°C or −80°C.

3.3.2 The extraction of DNA using a CTAB method (Day & Shattock 1997)

1. Homogenize 100 mg of frozen mycelium at −80°C in the presence of liquid nitrogen, and then add 800 μl of extraction buffer.
2. The resulting suspension is transferred to a microtubule volume of 2 ml and incubated in a water bath at 65°C for a period of 1 h. Every 15 minutes the contents mixed thoroughly. During this part of the extraction leads to the degradation of the cell walls and release the cell contents.
3. In each microtubules add 600 μl of chloroform and mix on the Vortex 10 seconds, and then centrifuged for 10 min at 13 000 rpm.
4. After centrifugation consists of two different phases. The upper layer, the approximate volume of about 500 μl, pipetted to a new microtube tube, then add 300 μl of isopropanol and incubated for 10 min at room temperature, and then centrifuged for 10 min at 13 000 rpm.
5. The supernatant is carefully poured away, and the residue was rinsed with 600 μl of 70% ethanol and centrifuged for 10 min at 13 000 rpm.
6. Decant the liquid phase, and the microtube dry at room temperature for 3 min or at 56°C in a heating block.
7. The resulting pellet resuspended in 50 μl TE buffer.
8. Isolation of DNA stored at −80°C.

<u>The composition of the buffer for the isolation</u>

100 mM Tris HCl, pH 8.0

1.4 M NaCl

20 mM EDTA
2% CTAB (cetyltrimethylammonium bromide) pH 8.0

The composition of TE buffer
10 mM Tris–HCl pH 8.0
1 mM EDTA

3.4 Application of conventional PCR

Reliable and sensitive method for the detection of pathogenic fungi from mycelium is implemented using universal primers ITS1/ITS4 (White et al. 1990). This pair of universal primers allows amplification and subsequent sequencing of the ITS region of ribosomal DNA of eukaryotes. ITS region is highly variable among morphologically different fungi species but it is conservative at the species level and in many genera of plant pathogenic fungi and it is used for phylogenetic analysis. ITS1/ITS4 is primers pair that can amplify the ITS region of eukaryotes and it can be used to check the success of DNA extraction. In addition, these primers are included in the survey in order to test their suitability for use within the protocol to sequencing identification. This method is proved to be reliable for the detection of various plant pathogenic fungi.

Prepare PCR master mix (reaction mixture) using 2x PCR master mix (ThermoFisher Scientific; http://www.thermofisher.com), which in its composition and quantity corresponds to the recommendations from the table.

Components	1 sample	No. of samples+[1]K^++nB+nR
RNase–free water	9 μl	
2x PCR master mix (ThermoFisher Scientific)	12.5 μl	
Primer 1 concentration of 10 μM	1.25 μl	
Primer 2 concetration of 10 μM	1.25 μl	
Sample (template DNA)	1 μl	
Total volume	25 μl	

Table 3: Quantities and components for prepareing the PCR mix in a volume of 25 μl.

[1]K+– positive control; nB – appropriate number of negative controls (Blank); nR – appropriate number of excess working reagent losses during manipulation; n= number of samples/10. (For every 10 samples to be tested, add one negative control (PCR mix with RNase free water) and capacity for another sample for the loss of reagents during manipulation. The amount of each component of PCR mix is multiplied by this number and added to the tube for the master mix)

Master mix is prepared in a separate, preferably sterile, in a laminar flow hood. The required ingredients are prepared, so that the solution of PCR mix and primers are melted, thoroughly mixed on vortex and then centrifuged briefly (10 seconds at 5000 rpm/min) in order to avoid localized differences in the concentration. In one microtube of 0.5 ml or more if necessary, pour in the measured quantities for the required number of samples and the negative controls following the recommendations below the table. All the ingredients except the DNA sample are poured. Thus prepared master mix is stirred on vortex and centrifuged briefly (10 seconds at 5000 rpm/min).

In prepared and marked 0.2 ml microtubes dispense by 24 μl mix, and then adds the extracted sample DNA and mixed by pipetting. Between samples must be changed extensions for micropipette. Microtubes is then centrifuged in order to equalize the concentration and placed in the Thermal cycler programmed according to the conditions shown in the table.

Conditions in Thermal cycler	Duration	t°C	Number of cycles
Initial denaturation	2–5 min	95°C	
Denaturation	0.5–1 min	94°C	25–40x
Annealing	0.5–1 min	50–68°C	
Extension	1 min	72°C	
Final extension	10 min	72°C	

Table 4: Thermal cycler conditions for PCR reactions.

As a positive reaction is considered to be the occurrence of the amplicon of about 500–600 bp, which must be present in the positive control (DNA extracted from reference isolates) and which should not be present in the B (blank – PCR mix with RNase free water) – negative control.

4 Analysis of PCR products

Whether applied in the study of plant pathogenic viruses or fungi and fungus–like organisms, after completing the PCR reaction it is necessary to make the results visible in order to be analyzed. The resulting product of amplification can be detected in several ways, but the most often is using agarose gel electrophoresis. Depending on the size of the PCR product, prepared gel with various concentrations of agarose (1–2%), stained with Ethidium bromide (EtBr) and examined under UV transilluminator (Lee et al. 2012).

If the size of the expected fragments is about 1000 bp, which is usually the case in the detection and characterization of viruses, and plant pathogenic fungi, a 1% agarose gel is used. The gel is loaded with all analyzed samples, all positive and negative controls as well as markers with fragments in the appropriate size range.

Preparation of the gel and gel electrophoresis (Lee et al. 2012)

1. Prepare 1% agarose gel by the addition of an appropriate mass of agarose in 1x TBE buffer.
2. Place and fix separators in the gel tray to create a mold. Also, put the combs into the gel mold to creat the wells.
3. Warm solution of agarose and buffer up to boil in a microwave (about 30 s, or until the agarose is melted).
4. Cool the melted agarose to a temperature of 50–60°C under the tap water and pour it in the gel mold with combs. When pouring the gel be careful do not create air bubbles.
5. Place the gel on the flat surface to be hardened and cooled (about 30 minutes).
6. When the gel is hardened, remove the combs and put it into electrophoresis with 1x TBE buffer.
7. From 20 μl of amplicon sample for electrophoresis is using 5 μl pre–mixed with colour. The resulting amplicons are colored that there could be monitoring their movement in the gel and to be heavier and thus sank to the bottom of the well.

Note: For colouring samples are used specially made or bought color Loading Day. Color is kept in the fridge, for longer periods kept in freez at −20°C.

8. Prepare samples for electrophoresis mixing with colour. On a piece of Parafilm make balls of approximately 1 μl loading color in the number of how many samples and one more for the marker. A marker is placed in an amount of 5 μl. Marker is kept in the freezer, and one that is used in the refrigerator.
9. Prepared samples are inflicting to wells required changing the extensions after each sample.
10. The electrophoresis is carried out under conditions of constant power of 100 V / 40 mA in a period of about 1 h or until the front of colour comes to about 1 cm before the bottom edge of the gel. Turn off the power source before releasing a gel from the device.
11. After completion of the electrophoresis gel to incubate for 15–30 minutes in a solution of Ethidium bromide (EtBr) in distilled water to a final concentration of 0.5 μg/ml. EtBr is a powerful mutagen and require special precautions in their work. Gel put into solution with Ethidium bromide using a wide spatula, container must be close. Gel leave for 15 minutes in order to reach the visualization of PCR products. When working with EtBr use 2 pairs of gloves, while the other pair touches only what is in constant contact with EtBr.

Note: An alternative method of staining the gel with the method in which a 0.1% solution of EtBr added directly to the dissolved gel (before pouring into the mold) in an amount of 5 μl to 100 ml of the gel. In this way, samples are immediately color while flowing through the agarose gel.

12. Flushing gel (bleaching) is carried out by incubation in distilled water for 15–20 min.
13. Observing the gel is carried out with UV transilluminator. DNA fragments are visible as strips of orange color. Gel expose UV–light as short as possible due to the fact that UV light gradually destroying and damages the DNA. UV light is also dangerous for the observers so be sure to use protection during observation.
14. Check the results by comparing the strip markers and reactions both positive and negative controls.
15. Take an image with camera with a yellow filter.
16. The gel inserts in a plastic bag that sticks and then into a container for destruction EtBr or leave it under UV light or direct sunlight until dries.

5 DNA cloning

Cloning of the DNA or recombinant DNA technology is one of the methods of molecular biology, which enables obtaining of identical copies of a specific DNA fragment or the gene of interest. Cloning is accomplished by incorporating the desired gene fragment into genom of plasmid or virus (phage) of certain bacteria, so that the inserted DNA sequence multiplies along with its replication. The genome of a carrier in which is incorporated a desired DNA fragment is referred to as a vector. Bacteria *Escherichia coli* is the most commonly used host which enables amplification of the vectors, although the vectors are designed for other types of bacteria and some simple eukaryotic cells such as yeast (Lodish et al. 2000).

In general, the DNA fragment that we want to clone is obtained using polymerase chain reaction (PCR). Subsequently, these fragments are combined with vector DNA and then inserted into a host organism which easy–to–grow, such as *E. coli* bacteria. This will generate a population of organisms in which recombinant DNA molecules are replicated along with the host DNA.

Although a very large number of host organisms and molecular cloning vectors are in use, the great majority of molecular cloning experiments begin with a pGEM®–T Easy plasmid vector and a laboratory strain DH5α of the bacterium *E. coli*. The pGEM®–T Easy vectors are linearized vectors with a single 3'–terminal thymidine at both ends which preventing recircularization of the vector and improve the efficiency of ligation of PCR product with adenine at both 3' ends generated by thermostable polymerase such as *GoTaq*, *Taq* and *AmpliTaq* as well as *Tfl* and *Tth* polymerase. In addition, the pGem–T easy vector containing the gene for bacterial resistance to ampicillin as well as the T7 and SP6 RNA polymerase promoter flanking a polylinker region (the multiple cloning region) within the lacZ gene coding region of the enzyme beta–galactosidase. Insertion of a DNA sequence into a plasmid leads to inactivation and unable to create a beta–galactosidase, and so that the colonies with a recombinant DNA molecule identified as white colonies on the medium containing lactose. The clones containing the plasmid with the PCR product produced white colonies, and plasmid clones with no PCR products give blue colonies on the medium with lactose because of the lacZ gene is functional and produces beta–galactosidase. Within polylinker region there are numerous restriction sites for different restriction enzymes (endonucleases) that allow the separation of the inserted DNA fragment from the plasmid (Lodish et al. 2000; Russell & Sambrook 2001).

The cloning of any DNA fragment essentially involves several steps: (1) amplification of DNA fragment that we want to clone, (2) ligation of PCR product with vector DNA (creation of recombinant DNA), (3) insertation of recombinant DNA into bacteria cells, (4) the reproduction of bacteria together with the recombinant DNA, (5) screening clones with recombinant DNA, and (6) plasmid extraction and purification from the bacterial cells.

5.1 Ligation of PCR product with vector

The use of pGEM–T Easy Vector Systems Kit (Promega GmbH, Mannheim, Germany; https://promega.com) based on using pGEMT Vectora which was linearized, containing one nucleotide thymine at the 3' end of both strands, has a gene responsible for resistance to amplicilin and a multiple cloning region that recognize a number of restriction enzymes. Ligation of the PCR product and the vector is simple and is based on the complementarity between adenine and thymine since most of the polymerase during synthesis of the fragment in a PCR reaction added one adenine nucleotide at the ends of DNA fragment.

Kit Contents:

1. pGEM®–T Vector (50 ng/μl)
2. Control Insert DNA (4 ng/μl)
3. T4 DNK Ligase
4. 2X Rapid Ligation Buffer, T4 DNK Ligase
5. JM109 Competent Cells, High Efficiency (optional, since there is posibility to purchase kit without the transformed cells)

Protocol (Anonymous, 2015b):

1. Thaw pGEM®–T vector, 2X Rapid Ligation buffer and control insert DNA, than vortex (it is especially important vortex the buffer vigorously because the salt in buffer can precipitated during store), spin down briefly and put on ice. Enzyme T4 DNA ligase is constantly kept on ice and no mix.

2. Vortex PCR product, spin down and put on ice.
3. Prepare mix according table 5.

Reaction component	Sample	Positive control
2X Rapid Ligation Buffer	5 µl	5µl
pGEM®–T or pGEM®–T Easy Vector (50ng)	1 µl	1 µl
PCR product	3 µl*	–
Control insert DNA	–	2 µl
T4 DNA Ligase (3 Weiss units/µl)	1 µl	1 µl
Nuclease free H_2O	up to a final volume of 10 µl	up to a final volume of 10 µl

Table 5: The amounts and components for the preparation of mix for application pGEM–T Easy Vector Systems kit.

* The amount of PCR product is determined depending on the concentration of the PCR product and the desired ratio of the vector and the PCR product. The ratio of the vector and the PCR product is from 8:1 to 1:8, but it is the best to use a ratio of from 3:1 to 1:3. The propriat amount of PCR product was calculated according to the following equation:

$$\frac{\text{ng of vector} \times \text{kb size of insert}}{\text{kb size of vector}} \times \text{PCR product:vector ratio} = \text{ng of PCR product}$$

Example: If ratio of the PCR product and vector is 3:1 and size of PCR fragment 1000 bp, than you should add 50 ng of the PCR product

$$\text{because } \frac{50 \text{ ng x } 1 \text{ kb}}{3 \text{ kb}} \times 3 = 50 \text{ ng}$$

4. In most cases you can use 3 ml of the PCR product. Make a mix containing all components except sample (PCR product), vortex and spin down briefly and then divide by 7 ml in tubes of 1.5 ml. After that in the tubes added to 3 ml of the PCR product, vortex, spin down and incubate overnight at 4°C (refrigerated).

5.2 Transformation of cell of Escherichia coli *strain DH5α*

Instead of transformed cells that are purchased in a kit, transformed cells of *E. coli* strain DH5α can be used. Transformation of *E. coli cells* includes opening of a pore in the cell wall of the bacteria and increase the number of receptor sites in order to put plasmid with the desired PCR product into bacterial cells.

Protocol (Gallitelli et al., Dipartimento di Scienze del suolo, della pianta e degli alimenti, Università degli Studi di Bari Aldo Moro, Bari, Italy, personal communication):

1. Inoculate bacteria cells in 3 ml of LB liquid media and incubate overnight at 37°C with constant stirring at 250 rpm (to do this in the same day as ligation of the PCR product).
2. Transfer the 50 ml suspension of bacteria in a new 10 ml of LB liquid media and incubate for 3.5 h at 37° C with constant stirring at 250 rpm (the bacteria suspension transfer in the new tube using large pipette or small pipette, but in this case, cut off the top of tips).
3. After incubation, centrifuge bacteria suspension at 4°C for 5 min at 6000 rpm.
4. Discard the supernatant and add 5 ml of transformation buffer on the opposite side of the pellet. Gently resuspend pellet tapping of the tube with your finger, incubate on ice for 20 min, and then centrifuge for 5 min at 6000 rpm at 4°C.
5. Discard the supernatant and add 600 μl of transformation buffer on the opposite side of the pellet. Gently resuspend pellet tapping of the tube with your finger and incubate on ice for 2 h.
6. After incubation, bacteria cells are transformed and prepared for cloning.

Transformation buffer

- 735 mg of $CaCL_2xH_20$ (necessary to dilate the pores of the cell wall)
- 5 mg of thymidine (active receptors of the bacterial cell)
- 1 ml of 1M Tris–HCl pH 7.5 (buffer function)

Make up the volume to 100 ml with sterile H_2O, and sterilize by bacteriological filter.

Instead of transformation buffer it can be used 100 mM $CaCl_2$ sterilized in the autoclave for 20 minutes.

5.3 Insertation of plasmid into transformed bacteria cells (Anonymous, 2015b)

1. Add 100 μl of transformed cells in 10 μl of ligation product and gently mix by tapping of the tube with your finger, and than incubate on ice for 30 min. Make sure you work on ice and the bacteria cell add using large pipette or small pipette, but in this case, cut off the top of tips. Before each addition of the bacteria cell, you must mix gently suspension of bacterial cells by tapping with your finger.
2. After that, incubate tube in water bath at 42°C for 3 min.

3. After this incubation, the tubes put on ice for 2 min, and than incubate at room temperature for additional 2 min.
4. After this incubation, in each tube add 300 μl of LB liquid media and incubate for 1 h at 37°C with constant stirring at 150 rpm.

5.4 Plating of bacteria cells (Anonymous, 2015b)

1. Thaw LB media with agar and cool down at 50°C, and then add 1 ml of ampicillin concentration of 100 mg/ml per 1 ml of the media.
2. Spilled the media in Petri dishes and allow tightening (for each sample prepared by 2 Petri dishes).
3. Distribute carefully 40 μl of X–Gal onto Petri dishes 15 min before planting of bacteria cell and leave Petri dishes to absorb X–Gal (X–Gal applied by pipette as drops and then evenly distributed on the media).
4. Aftre that, add 100 μl of transformed bacteria cells by pipette as drops and distribute well onto the media (be sure to use a smaller pipette but cut off the top of tips).
5. Spin–down the remaining of the bacteria suspension, discard 100 μl of the supernatant and gently resuspend pellet in the remaining volume by tapping with your finger.
6. Add 100 μl of bacteria suspension onto second Petri dishes and distribute well onto the media.
7. Incubate Petri dishes over night at 37°C, so the lid downwards.
8. After incubation, appears white (which contain plasmid) and blue colonies (which do not contain plasmid).
9. Petri dishes kept in the refrigerator until the moment of purification of plasmid.

X–Gal

Dissolve 100 mg of 5–bromo–4–chloro–3–indolyl–β–D–galactopyranoside in 2 ml of 2 NN dimethyformamide and store at –20°C.

5.5 Plasmid DNA extraction from Escherichia coli *(Gallitelli et al., Dipartimento di Scienze del suolo, della pianta e degli alimenti, Università degli Studi di Bari Aldo Moro, Bari, Italy, personal communication)*

1. Inoculate 3 ml LB media containing 100 μg/ml ampicilin with one white colony picked up from Petri dishes and incubate at 37°C with constant stirring at 250 rpm.
2. After incubation, transfer part of suspension in a 1.5 ml tube and centrifuge for 30 s at 14000 rpm. Discard the supernatant and dry tube on the filter paper. Repeated this step with the rest of suspension.

3. Add 350 μl of STET solution in each tube (8% sucrose, 20 mM Tris–HCl pH 8, 50 mM EDTA, 0.5% Triton) and 15 μl of 20 μm/μl lysozyme and dissolve pellet on vortex.
4. Incubate tubes on boiling water for 40 s and immediately put on ice and incubate for 5 min.
5. After incubation, samples centrifuge for 20 min at 14000 rpm.
6. After centrifugation, remove the bacterial chromosome DNA with a sterile toohtpick.
7. Add 175 μl of phenol and 175 μl of chloroform: isoamyl alcohol mixture in 24:1 proportion, vortex for 40 s and centrifuge for 10 min at 14000 rpm.
8. Recover the aqueous phase in new 1.5 ml tube and add 200 μl of 5 M ammonium acetate pH 5.5 and 1 ml of cold absolte ethyl alcohol, and vortex.
9. Centrifuge tubes at 14000 rpm for 10 min, and than discard the supernatant and dry well tube on the filter paper.
10. Wash the pellet with 500 μl of cold 70% ethyl alcohol, and than centrifuge for 3 min at 14000 rpm. Discard supernatant by pipette and dry pellet at 65°C for 5 min.
11. Resuspend pellet in 50 μl of TE buffer (10 mM Tris–HCl pH 8, 1mM EDTA) and digest RNA with 1 μl of RNaseA (concentration 10 mg/ml) at 37°C for 30 min.
12. Add 30 μl of PEG–NaCl (20% Polyetilen glikol 6000 and 2.5 M NaCl) in each tube and mix carefully on vortex. Incubate samples on ice for 1 h.
13. After incubation, centrifuge tubes for 10 min at 14000 rpm.
14. Discard supernatant and wash pellet with 300 μl of cold 70% ethyl alcohol.
15. Centrifuge tubes for 2 min at 14000 rpm, discard supernatant and dry pellet at 65°C for 5 min.
16. After that, dissolve pellet in 30 μl of RNase free water.

The pellet represents the cloned target DNA fragment wich was obtained in a large amount and can be used for further analizis using other methods, such as, for example, sequencing to verify the identity or other molecular methods for identification of phytopathogenic viruses, fungi and fungus–like organisms.

References

Anderson, P. K., Cunningham, A. A., Patel, N. G., Morales, F. J., Epstein, P. R. & Daszak, P. (2004). Emerging infectious diseases of plants: pathogen pollution, climate change and agrotechnology drivers. *Trends in Ecology & Evolution 19*, 535–544. DOI: https://doi.org/10.1016/j.tree.2004.07.021

Anonymous (2008). RNAqueous® Kit (Part Number AM1912), Protocol. https://tools.thermofisher.com/content/sfs/manuals/cms_055306.pdf

Anonymous (2012a). Qiagen® OneStep RT–PCR Handbook. https://www.qiagen.com/us/resources/download.aspx?id=57743726-84e1-423a-9d8f-a3fa89bbe7eb&lang=en

Anonymous (2012b). RNeasy® Mini Handbook, Fourth Edition. http://www.bea.ki.se/documents/EN–RNeasy%20handbook.pdf

Anonymous (2015a). DNeasy® Plant Handbook. https://www.researchgate.net/file.PostFileLoader.html?id=56ff54f2eeae39355963f1b1&assetKey=AS%3A346292619366400%401459574002159

Anonymous (2015b). pGEM®–T and pGEM®–T Easy Vector Systems, Technical Manual. https://www.promega.com/-/media/files/resources/protocols/technical-manuals/0/pgem-t-and-pgem-t-easy-vector-systems-protocol.pdf.

Bekesiova, I., Nap, J. P. & Mlynarova, L. (1999). Isolation of high quality DNA and RNA from leaves of the carnivorous plant *Drosera rotundifolia*. *Plant Molecular Biology Reporter 17*, 269–27. DOI: https://doi.org/10.1023/A:1007627509824.

Day, J. P. & Shattock, R. C. (1997). Aggressiveness and other factors relating to displacement of population of *Phytophthora infestans* in England and Wales. *European Journal of Plant Patholoogy 103*, 379–391. DOI: https://doi.org/ 10.1023/A:1008630522139.

Iandolino, A. B., Goesda Silva, F., Lim, H., Choi, H., Williams, L. E. & Cook, D. R. (2004). High–quality RNA, cDNA, and derived EST libraries from grapevine (*Vitis vinifera* L.). *Plant Molecular Biology Reporter 22*, 269–278. DOI: https://doi.org/10.1007/BF02773137

Konstantinova, P., Bonants, P. J. M., van Gent–Pelzer, M. P. E., van der Zouwen, P. & van den Bulk, R. (2002). Development of specific primers for detection and identification of *Alternaria* spp. in carrot material by PCR and comparation with blotter and plating assays. *Mycological Research 106*, 23–33. DOI: https://doi.org/10.1017/S0953756201005160.

Kroon, L. P. N. M., Verstappen, E. C. P., Kox, L. F. F., Flier, W. G. & Bonants, P. J. M. (2004). A Rapid Diagnostic Test to Distinguish Between American and European Populations of *Phytophthora ramorum*. *Phytopathology 94*, 613–620. DOI: https://doi.org/10.1094/PHYTO.2004.94.6.613.

Krstić, B., Bulajić, A. & Đekić, I. (2008). *Tomato spotted wiltvirus*, TSWV–Standardna operativna procedura za fitopatološke dijagnostičke laboratorije. Univerzitet u Beogradu–Poljoprivredni fakultet i Ministarstvo poljoprivrede, vodoprivrede i šumarstva, Beograd.

Krstić, B., Bulajić, A., Ivanović, M., Stanković, I. & Vučurović, A. (2010). *Alfalfa mosaicvirus*, AMV–Standardna operativna procedura za fitopatološke dijagnostičke laboratorije. Univerzitet u Beogradu–Poljoprivredni fakultet i Ministarstvo poljoprivrede, vodoprivrede i šumarstva, Beograd.

Lee, P. Y., Costumbrado, J., Hsu, C. Y. & Kim, Y. H. (2012). Agarose Gel Electrophoresis for the Separation of DNA Fragments. *Journal of Visualized Experiments 62*, e3923. DOI: https://doi.org/10.3791/3923

Lodish, H., Berk, A., Zipursky, S. L., Matsudaira, P., Baltimore, D. & Darnell, J. (2000). *Molecular cell biology*, 4th edition. New York: W. H. Freeman, DOI: https://doi.org/10.1016/S1470–8175(01)00023–6

McKirdy, S. J., Washer, S., Berryman, D., Sargent, K., Selladurai, S., Jones, R. A. C. & Coutts, B. (1998). Virus testing in chickpea and lentil seed. Pea seed–borne mosaic and other viruses. Pulse Research and Industry Development in Western Australia, 1998. Updates, Observation City, pp. 94–95. https://doi.org/10.1094/PDIS–11–15–1249–RE

McPherson, M. J. & Møller, S. G. (2000). *PCR*. BIOS Scientific Publishers Ltd.

Miller, S. A., Beed, F. D. & Harmon, C. L. (2009). Plant disease diagnostic capabilities and networks. *Annual Review of Phytopathology 47*, 15–38. DOI: https://doi.org/10.1146/annurev–phyto–080508–081743

Reece, R. J. (2004): Analysis of genes and genomes. John Wiley & Sons, ISBN–10: 1449635962

Russell, D. W. & Sambrook, J. (2001). Molecular cloning: a laboratory manual. Cold Spring Harbor, N.Y: Cold Spring Harbor Laboratory, https://doi.org/10.1086/394015

Sellner, L. N. & Turbett, G. R. (1998). Comparison of three RT–PCR methods. *Biotechniques 25,* 230–234.

Strange, R. N. & Scott, P. R. (2005). Plant disease: a threat to global food security. *Annual Review of Phytopathology 43*, 83–116. DOI: https://doi.org/10.1146/annurev.phyto.43.113004.133839

White, T. J., Bruns, T., Lee, S. & Taylor, J. (1990). Amplification and direct sequencing of fungal ribosomal RNA genes for phylogenetics. In: M. Innis, D. Gelfand, J. Sninsky, T. White (eds.), San Diego Academic Press, Inc., *PCR protocols: A guide to methods and applications*, pp. 315–322. DOI: https://doi.org/10.1016/B978–0–12–372180–8.50042–1.

Zeng, Y. & Yang, T. (2002). RNA Isolation From Highly Viscous Samples Rich in Polyphenols and Polysaccharides. *Plant Molecular Biology Reporter 20,* 417a–417e. DOI: https://doi.org/10.1007/BF02772130

Real-time PCR detection of quarantine plant pathogenic bacteria in potato tubers and olive plants

Milan Ivanović, Nemanja Kuzmanović, Nevena Zlatković

Abstract

This chapter focuses on molecular detection of quarantine plant pathogenic bacteria associated with potato and olive plants. A real-time PCR for detection of two bacteria of potato in Europe: *Ralstonia solanacearum* race 3 and *Clavibacter michiganensis* subsp. *sepedonicus*, is described. This method allows the simultaneous detection of both species in a single PCR reaction with an internal control from potato. Described protocol is sensitive and specific and can be used in large scale screening tests. Quarantine pest *Xylella fastidiosa* is recently detected in Europe for the first time. Considering the importance of early detection, procedure for real-time PCR detection of *Xylella fastidiosa* in olive plant tissue is described.

Simultaneous detection of *Ralstonia solanacearum* race 3 and *Clavibacter michiganensis* subsp. *sepedonicus* in potato tubers by a multiplex real-time PCR assay

How to cite this book chapter:
Ivanović, M., Kuzmanović, N. and Zlatković, N. 2019. Real-time PCR detection of quarantine plant pathogenic bacteria in potato tubers and olive plants. In: Vucelić Radović, B., Lazić, D. and Nikšić, M. (eds.) *Application of Molecular Methods and Raman Microscopy/Spectroscopy in Agricultural Sciences and Food Technology*, Pp. 83–95. London: Ubiquity Press. DOI: https://doi.org/10.5334/bbj.f. License: CC-BY 4.0

1 Introduction

Ralstonia solanacearum (Smith) Yabuuchi et al. race 3 (Rs) and *Clavibacter michiganensis* (Smith) Davis et al. subsp. *sepedonicus* (Spieckermann et Kotthoff) Davis et al., (Cms) are the causal agents of brown-rot (Figure 1) and ring-rot (Figure 2) of potato, respectively. These diseases represent a serious threat to potato (*Solanum tuberosum*) production in temperate climates. Both bacteria are listed as A2 pests in the EPPO region and as zero-tolerance quarantine organisms in the European Union. These bacteria remain latent for a long time in asymptomatic potato tubers which are one of the main factors for the disease's dissemination. The existing phytosanitary regulations rely on the availability of pathogen-free seed tubers.

2 Materials, Methods and Notes

Since the protocols involve detection of quarantine organisms and include the use of viable cultures of Rs andCms, it is necessary to perform the procedures under suitable quarantined conditions with adequate waste disposal facilities and under the conditions of appropriate licenses as issued by the official plant quarantine authorities.

2.1 Sample preparation – potato tubers

Note:

- The standard sample size is 200 tubers per test. Larger numbers of tubers in the sample will lead to inhibition or difficult interpretation of the results. However, the procedure can be conveniently applied for samples with less than 200 tubers where fewer tubers are available.
- Detection methods described below are based on testing of samples of 200 tubers.
- Optional pre-treatment in advance to sample preparation: wash the tubers. Use appropriate disinfectants (chlorine compounds when PCR-test is to be used in order to remove eventual pathogen DNA) and detergents between each sample. Air-dry the tubers.
- This washing procedure is particularly useful (but not required) for samples with excess soil and if a PCR-test or direct isolation procedure is to be performed.

2.1.1. Remove with a clean and disinfected scalpel or vegetable knife the skin at the heel end of each tuber so that the vascular tissue becomes visible. Carefully cut out a small core of vascular tissue at the heel end and keep the amount of non-vascular tissue to a minimum.

Figure 1: Symptoms of brown rot on stored potato tubers caused by bacteria *Ralstonia solanacearum*. Note the brown staining of the vascular ring. (Foto: M. Ivanović).

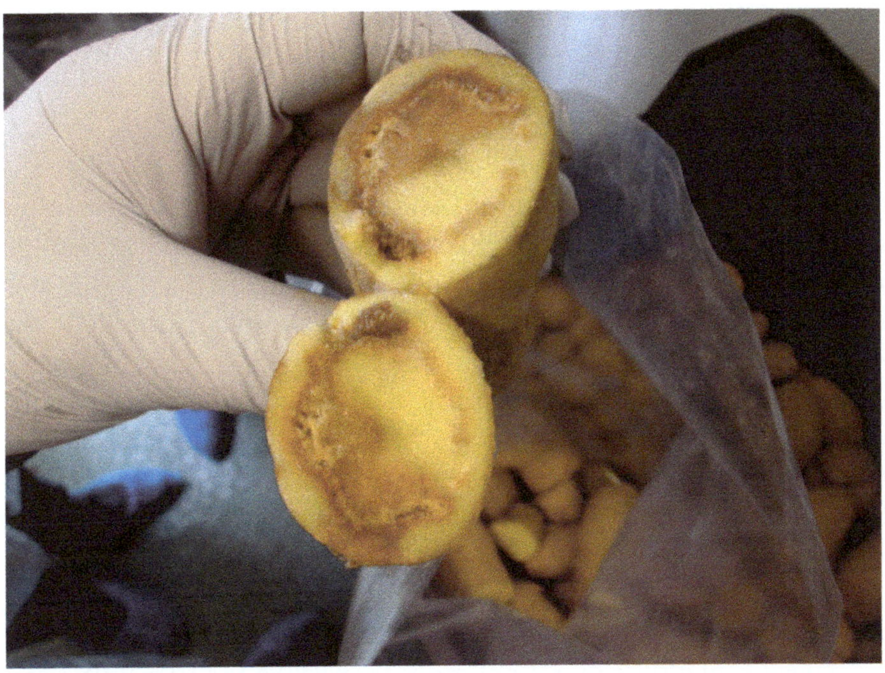

Figure 2: Breakdown and hollowing of stored potato tuber with ring rot, caused by bacteria *Clavibacter michiganensis*. subsp. *sepedonicus* (Photo: M. Ivanović).

Note: If during removal of the heel end core suspect symptoms of ring rot are observed, the tuber should be visually inspected after cutting near the heel end. Any cut tuber with suspected symptoms should be suberised at room temperature for two days and stored under quarantine (at 4 to 10°C) until all tests have been completed.

2.1.2. Collect the heel end cores in unused disposable containers which can be closed and/or sealed (in case containers are reused they should be thoroughly cleaned and disinfected using chlorine compounds). Preferably, the heel end cores should be processed immediately. If this is not possible, store them in the container, without addition of buffer, refrigerated for not longer than 72 hours or for not longer than 24 hours at room temperature. Drying and suberisation of cores and growth of saprophytes during storage may hinder detection of the brown rot and ring rot bacterium.

2.1.3. Process the heel end cores by one of the following procedures: either,

(a) cover the cores with sufficient volume (approximately 40 ml) of extraction buffer (see recipe below) and agitate on a rotary shaker (50 to 100 rpm) for four hours below 24°C or for 16 to 24 hours refrigerated; or

(b) homogenize the cores with sufficient volume (approximately 40 ml) of extraction buffer, either in a blender (e.g. Waring or Ultra Thurax) or by crushing in a sealed disposable maceration bag (e.g. Stomacher or Bioreba strong gauge polythene, 150 mm × 250 mm; radiation sterilized) using a rubber mallet or suitable grinding apparatus (e.g. Homex, Bioreba).

Note:

- The risk of cross-contamination of samples is high when samples are homogenized using a blender. Take precautions to avoid aerosol generation or spillage during the extraction process. Ensure that freshly sterilized blender blades and vessels are used for each sample. If the PCR test is to be used, avoid carry-over of DNA on containers or grinding apparatus. Crushing in disposable bags and use of disposable tubes is recommended where PCR is to be used.
- Recipe for extraction buffer (50 mM phosphate buffer): Na_2HPO_4 (anhydrous), 4.26 g; KH_2PO_4, 2.72 g; distilled water, 1 L. Dissolve ingredients, adjust pH to 7.0 and sterilize by autoclaving at 121°C for 15 min.

2.1.4. Decant the supernatant. If excessively cloudy, clarify either by slow speed centrifugation (at not more than 180 g for 10 minutes at a temperature between 4 to 10°C) or by vacuum filtration (40 to 100 μm), washing the filter with additional (10 ml) extraction buffer.

2.1.5. Concentrate the bacterial fraction by centrifugation at 7 000 g for 15 minutes (or 10 000 g for 10 minutes) at a temperature between 4 to 10°C and discard the supernatant without disturbing the pellet.

2.1.6. Resuspend the pellet in 1,5 ml pellet buffer (see recipe below). Use 500 μl to test for Rs, 500 μl for Cms, and 500 μl for reference purposes. Add sterile glycerol to final concentration of 10 to 25 % (v/v) to the 500 μl of the reference aliquot and to the remaining test aliquot, vortex and store at – 16 to –24°C (weeks) or at –68 to –86°C (months). Preserve the test aliquots at 4 to 10°C during testing. Repeated freezing and thawing is not advisable. If transport of the extract is required, ensure delivery in a cool box within 24 to 48h.

Note:

- Recipe for pellet buffer (10 mM phosphate buffer): $Na_2HPO_4{\bullet}12H_2O$, 2.7 g; $NaH_2PO_4{\bullet}2H_2O$, 0.4 g; distilled water, 1 L. Dissolve ingredients, adjust pH to 7.2 and sterilize by autoclaving at 121°C for 15 min.
- It is imperative that all Rs and Cms positive controls, and samples are treated separately to avoid contamination.

2.2 DNA extraction (method according to Pastrik 2000)

1.1.1. Pipette 220 μl of lysis buffer (100 mM NaCl, 10 mM Tris-HCl [pH 8.0], 1 mM EDTA [pH 8.0]) into a 1.5 ml Eppendorf tube.

1.1.2. Add 100 μl sample extract and place in a heating block or water bath at 95°C for 10 min.

1.1.3. Put tube on ice for 5 min.

1.1.4. Add 80 μl Lysozyme stock solution (50 mg Lysozyme per ml in 10 mM Tris HCl, pH 8,0) and incubate at 37°C for 30 min.

1.1.5. Add 220 μl of Easy DNA® solution A (Invitrogen), mix well by vortexing and incubate at 65°C for 30 min.

1.1.6. Add 100 μl of Easy DNA® solution B (Invitrogen), vortex vigorously until the precipitate runs freely in the tube and the sample is uniformly viscous.

1.1.7. Add 500 μl of chloroform and vortex until the viscosity decreases and the mixture is homogeneous.

1.1.8. Centrifuge at 15 000 g for 20 min at 4°C to separate phases and form the interphase.

1.1.9. Transfer the upper phase into a fresh Eppendorf tube.

1.1.10. Add 1 ml of 100% ethanol (–20°C) vortex briefly and incubate on ice for 10 min.

1.1.11. Centrifuge at 15 000 g for 20 min at 4°C and remove ethanol from pellet.

1.1.12. Add 500 µl 80% ethanol (–20°C) and mix by inverting the tube.
1.1.13. Centrifuge at 15 000 g for 10 min at 4°C, save the pellet and remove ethanol.
1.1.14. Allow the pellet to dry in air or in a DNA speed vac.
1.1.15. Resuspend the pellet in 100 µl sterile UPW and leave at room temperature for at least 20 minutes.
1.1.16. Store at –20°C until required for PCR.
1.1.17. Spin down any white precipitate by centrifugation and use 5 µl of the supernatant containing DNA for the PCR.

Note:

- It is also recommended to prepare one decimal dilution of sample DNA extract (1:10 in sterile distilled water) for PCR analysis.
- Other DNA extraction methods, e.g. Qiagen DNeasy Plant Kit, could be applied providing that they are proven to be equally as effective in purifying DNA from control samples containing 10^3 to 10^4 pathogen cells per ml.

2.3 Real-time PCR assay (method according to Massart et al. 2014)

This multiplex real-time PCR assay allows simultaneous detection of Rs and Cms in potato tubers. For both bacteria, the primers and probes (Table 1) were selected in the rRNA gene intergenic spacer sequences. Additionally, the reliability of this molecular diagnostic test has been improved by the simultaneous amplification of an internal control, corresponding to a potato gene co-extracted from the sample. For the internal control, primers and probes (Table 1) were designed based on chloroplastic ATP synthase beta-subunit from *Solanum tuberosum*. The Minor Groove Binder (MGB) probes were supplied by Applied Biosystem with a 5' covalently attached reporter dye (FAM, VIC or NED), a nonfluorescent quencher and MGB moiety at the 3' end. The composition of reaction mix and thermal cycling conditions are given in Tables 2 and 3.

The proper negative and positive controls are essential for eliminating false-negative or false-positive results. In this regard, the following negative controls should be included in the real-time PCR test:

- DNA extracted from sample extract that was previously tested negative for Rs and Cms. Sample extracts should be as free as possible from soil. It could therefore, in certain cases, be advisable to prepare extracts from washed potatoes.
- Buffer controls used for extracting the bacterium and the DNA from the sample,
- Incorporate a negative control sample containing only PCR reaction mix and add the same source of nuclease-free water as used in the PCR mix in place of sample.

Primer or probe[a]	Sequence (5'-3')	Dye
MultiRaso-F	CGCGGAGCATTGATGAGAT	
MultiRaso-R	TCGTAATACTGGTTGATACAATCACAAC	
MultiRaso-P	CTCGCAAAAACGC	VIC
MultiClav-F	TGGTTTCTTGTCGGACCCTTT	
MultiClav-R	CGTCCACTGTGTAGTTCTCAATATACG	
MultiClav-P	CGTCGTCCCTTGAGTGG	FAM
MultiPot-F	GGTTTCGTAATGTTCCTCACCAA	
MultiPot-R	AAAGGTATTTATCCAGCAGTAGATCCTT	
MultiPot-P	CATGGTTGACGTTGAAT	NED

Table 1: Primers and probes for quantitative real-time PCR.

[a]F, forward; R, reverse; P, probe

Reagent	Volume
Qiagen mix	12.5 µl
Molecular grade water	0.75 µl
10 µM Forward MultiPot-F Primer	0.75 µl
10 µM Reverse MultiPot-R Primer	0.75 µl
10 µM TaqMan MultiPot-P Probe	1.25 µl
10 µM Forward MultiRaso-F Primer	0.75 µl
10 µM Reverse MultiRaso-R Primer	0.75 µl
10 µM TaqMan MultiRaso-P Probe	0.5 µl
10 µM Forward MultiClav-F Primer	0.75 µl
10 µM Reverse MultiClav-R Primer	0.75 µl
10 µM TaqMan MultClav-P Probe	0.5 µl
Template DNA	5 µl
Total	**25 µl**

Table 2: Reaction mix for quantitative real-time PCR.

95°C 15 min	1 cycle
95°C 20 sec	40 cycles
60°C 60 sec	

Table 3: Real-time PCR conditions.

In addition, the following positive controls should be also included:

- DNA extracted from sample extract that was previously tested negative for Rs and Cms spiked withsuspensions of Rs and Cms(several dilutions)
- DNA extracted from suspension of 10^6 cells per ml of Rsand Cms in water from a virulent reference strain (e.g. NCPPB 4156 = PD 2762 = CFBP 3857 for Rs; NCPPB 2140 or NCPPB 4053 for Cms).
- If possible use also DNA extracted from positive control samples in the PCR test.

To avoid potential contamination prepare positive controls in a separate environment from samples to be tested.

3 Acknowledgements

This work was funded by EU Commission project AREA, no. 316004. We would like to thank Dutch General Inspection Service and NAK Institute in Emmeloord, especially Miriam Kooman and Jaap Janse for arranging our visit, Robert Vreeburg and Robert Bollema for laboratory support.

4 References

Massart, S., Nagy, C., & Jijakli, M. H. (2014). Development of the simultaneous detection of *Ralstonia solanacearum* race 3 and *Clavibacter michiganensis* subsp. *sepedonicus* in potato tubers by a multiplex real-time PCR assay. *European Journal of Plant Pathology*, 138(1), 29–37. DOI: https://doi.org/10.1007/s10658-013-0294-4

Pastrik, K. H. (2000). Detection of *Clavibacter michiganensis* subsp. *sepedonicus* in potato tubers by multiplex PCR with coamplification of host DNA. *European Journal of Plant Pathology*, 106(2), 155–165. DOI: https://doi.org/10.1023/A:1008736017029

Real-Time PCR detection of *Xylella fastidiosa* subsp. *pauca* (CoDiRo strain) from olive plants

1 Introduction

The olive quick decline syndrome (OQDS) is a disease that appeared suddenly a few years ago in the province of Lecce (Italy). In 2013, it has been found that the most relevant factor for this disease is a quarantine pathogen *Xylella fastidiosa*. This was the first confirmed record in the European Union. In addition, almond, oleander, cherry and several other perennial ornamentals have been reported as hosts (Cariddi et al. 2014, EPPO 2016). Isolation and culturing of the bacterium on media are fundamental in phytobacteriology, but considering that some *Xylella* subspecies are very slow-growth, molecular and serological techniques showed as more suitable methods for screening a large number of samples. The purpose of this manuscript is to describe procedure for real-time PCR detection of *X. fastidiosa* in plant tissue.

2 Materials, Methods and Notes

Since the protocols involve detection of a quarantine organisms and include the use of viable cultures of *X. fastidiosa*, it is necessary to perform the procedures under suitable quarantined conditions with adequate waste disposal facilities and under the conditions of appropriate licenses as issued by the official plant quarantine authorities.

2.1 Collecting samples

During the training, we collected mostly symptomatic or asymptomatic olive plant material for *X. fastidiosa* isolation. Typical symptoms for OQDS are the presence of leaf scorch (Figure 1) and scattered desiccation of twigs and small branches. In the early stages of the infection, symptoms prevail on the upper part of the canopy. Later, these symptoms become increasingly severe and progress into the rest of the crown, which becomes blighted.

2.2 Sample preparation

Extraction of *X. fastidiosa* DNA from culture and plant tissue for molecular analyses has been achieved by both standard commercial column kits and

Figure 1: First symptoms of OQDS on olive leaves (Photo: N. Zlatković).

by basic CTAB buffer. Basal leaf portion and peduncles excised from mature leaves in total weight between 0,5–0,8 g are used for DNA extraction. Selected leaves should be representative of the whole sample. Symptomatic leaves have priority.

2.3 CTAB-based total nucleic acid extraction from plant tissue

2.3.1. Weigh out 0,5–0,8 g of fresh small pieces of midribs and petioles (1/4 if lyophilized), transfer the tissue into the extraction bags and add 2ml of CTAB. Crush with a hammer and homogenize.
2.3.2. In each extraction bag add 3ml of CTAB.
2.3.3. Transfer 1ml of sap into a 2ml microcentrifuge tube.
2.3.4. Heat samples at 65°C for 30 minutes.
2.3.5. Centrifuge samples at 10,000 rpm for 5 minutes and transfer 1ml to a new 2ml microcentrifuge tube, being careful not to transfer any of the plant tissue debris. Add 1ml of Chloroform: Isoamyl Alcohol 24:1 and mix well by shaking or vortex.
2.3.6. Centrifuge sample at 13,000 rpm for 10 minutes. Transfer 750 ml to a 1.5 ml microcentrifuge tube and add 450 μl (approximately 0.6 volume of cold 2-Propanol. Mix by inverting 2 times. Incubate at 4°C or –20°C for 20 minutes.
2.3.7. Centrifuge the samples at 13.000 rpm for 20 minutes and decant the supernatant.
2.3.8. Wash pellet with 1ml of 70% ethanol.
2.3.9. Centrifuge sample at 13,000 rpm for 10 minutes and decant 70% ethanol.
2.3.10. Air-dry the samples or use the vacuum.
2.3.11. Re-suspend the pellet in 100μl of TE or RNAse- and DNase-free water.
2.3.12. Extracts of total nucleic acid can be stored at 4° C for immediate use or at 2.3.13. 20°C for use in the future.
2.3.14. Determine the concentration at the spectrophotometer (Nanodrop 1000 or similar). Read the absorption (A) at 260nm and at 280 nm. Optimal A260/280 ratio should be close to 2 for high quality nucleic acid.
2.3.15. Adjust the concentration to 50–100ng/μ l, and use 2 μl (in a final volume of 20–25μl) to set up the conventional and real time PCR reactions.

Note: Recipe for CTAB buffer: 2% CTAB (Hexadecyl trimethyl-ammonium bromide), autoclaved 0.1M TrisHCl pH 8, autoclaved 20mM EDTA, autoclaved 1.4M NaCl, 1% PVP-40.

2.4 DNA extraction using commercial kit

DNeasy Plant Mini Kit, Cat. No. 69104 – Qiagen, Valencia, CA

2.4.1. Weigh out 200 mg fresh tissue (1/4 if lyophilized) and homogenize with mortar and pestle in liquid nitrogen and transfer powered tissue into 2ml microcentrifuge tubes. Remaining tissues can be stored at –20°C for future use.
2.4.2. Add 800 µl of the Qiagen DNeasy Plant Mini extraction kit AP1 buffer and 8 µl of RNase A stock solution (100 mg/ml) into a sample tube.
2.4.3. Incubate cellular lysate at 65°C for 10 min.
2.4.4. Add 260 µl of Buffer AP2 to the lysate, vortex briefly and incubate on ice for 5 min.
2.4.5. Centrifuge at 20,000 × g (14,000 rpm) for 10 min.
2.4.6. Pipet lysate into a QIAshredder Mini Spin Column (lilac colored column) in a 2 ml collection tube and centrifuge for 2 min at 20,000 x g (14,000 rpm), then, discard the column (typically about 500 µl of lysate can be recovered).
2.4.7. Measure the volume and add 1.5 volumes of Buffer AP3/E to the lysate and mix by pipetting.
2.4.8. Transfer 650 µl of the mixture including any precipitate to the DNeasy Mini Spin Column sitting in a 2 ml collection tube. Centrifuge at 6000 × g (8000rpm) for 1 min. (Discard flow through).
2.4.9. Repeat Step 12 with the remaining portion of the mixture. Discard flow-through and collection tube.
2.4.10. Place the spin column in a new 2 ml collection tube. Add 500 µl of Buffer AW to the column and centrifuge at 8000 rpm for 1 min. Discard flow-through.
2.4.11. Add another 500 ml of AW and centrifuge for 2 min at 20,000 x g (14,000 rpm) to dry the membrane.
2.4.12. Transfer the spin column to a 1.5 ml microcentrifuge tube and pipet 200 µl of Buffer AE (room temperature) onto the column membrane. Incubate for 5 min at room temperature and then centrifuge for 1 min at 6,000 x g (8000rpm) to collect DNA elution (do not allow the column to dry).
2.4.13. Extracts of total genomic DNA can be stored at 4° C for immediate use or at –20°C for use in the future.

2.5 Real-time PCR (method according to Harper et al., 2010)

Harper et al. (2010) developed Real-time PCR assays targeted to the rimM gene of *X. fastidiosa*, which detected all bacterial subspecies. The primer set has been previously tested and proved to be suitable for detection of CoDiRo strain in olive tissues (Table 1). The composition of reaction mix and thermal cycling conditions are given in Tables 2 and 3. Each reaction should include the positive, the negative and the non-template controls. For this method, samples should be run in duplicate wells. Presence of DNA band of expected size means

Primer or probe	Sequence (5'-3')
XF-F (forward)	CACGGCTGGTAACGGAAG
XF-R (reverse)	GGGTTGCGTGGTGAAATCAAG
XF-P (probe)	6FAM-TCGCATCCCGTGGCTCAGTCC-BHQ1

Table 1: Primers and probes for quantitative real-time PCR.

Reagent	Volume
Total genomic DNA	1 μl
2× master mix for probes	5.5 μl
10 μM Forward Primer	0.3 μl
10 μM Reverse Primer	0.3 μl
10 μM TaqMan Probe	0.1 μl
Molecular grade water	3.8 μl
Total	**11 μl**

Table 2: Reaction mix for quantitative real-time PCR.

50°C 2 min	1 cycle
95°C 10 min	1 cycle
94°C 10 sec	39 cycles
62°C 40 sec	

Table 3: Real-time PCR conditions.

that sample is positive. If a sample produces a FAM Cq value in the range of 0.00< FAM Cq< 35.00, the sample is determined to be positive for *X. fastidiosa* and if produces a FAM Cq=0.00 or >35.0, then it is determined to be negative. If the FAM Cq value is between 32.01 and 34.99, then the samples have to be tested again in real-time PCR to confirm the result.

3 Acknowledgements

This work was funded by EU Commission project AREA, no. 316004. I would like to thank National Research Council (CNR) and Institute for Sustainable Plant Protection (IPSP) in Bari, especially Dr Maria Saponari who was excellent supervisor. All detection protocols are published by CNR Instituto per la Protezione Sosteinibile delle Piante (Bari, Italy), in the framework of workshop "Current tools for the detection of *Xylella fastidiosa* in host plants and vectors" (Saponari et al. 2014).

4 References

Cariddi, C., Saponari, M., Boscia,D., Stradis, A. D., Loconsole, G., Nigro, F., Porcelli, F., Potere, O., & Martelli, G. P. (2014). Isolation of a *Xylella fastidiosa* strain infecting olive and oleander in Apulia, Italy. *Journal of Plant Pathology*, 96(3), 1–5. DOI: https://doi.org/10.4454/JPP.V96I2.024

EPPO (2016). First report of *Xylella fastidiosa* in EPPO region, Special Alert. Retrieved from http://www.eppo.int/QUARANTINE/special_topics/Xylella_fastidiosa/Xylella_fastidiosa.htm

Harper, S. J., Ward, L. I., & Clover, G. R. G. (2010). Development of LAMP and Real-Time PCR Methods for the Rapid Detection of Xylella fastidiosa for Quarantine and Field Applications. *Phytopathology*, 100(12), 1282–1288, DOI: https://doi.org/10.1094/PHYTO-06-10-0168

Saponari, M., Loconsole, G., Potere,O., Palmisano, F., & Boscia, D. (2014). Current protocols for detection of *Xylella fastidiosa* in host plants and vectors. Workshop manual, CNR Instituto per la Protezione Sosteinibile delle Piante, Bari, Italy. Retrieved from ftp://ftpfiler.to.cnr.it:21001/Xylella_symposium/Workshop%20manuals/WORKSHOP%20MANUAL%20DETECTION%20ENG.pdf

Application of PCR in Food Biochemistry

Milica Pavlićević and Biljana Vucelić-Radović

Abstract

For detection of genetically modified organisms, several different methodologies could be employed. Different types of quantitative PCR (qPCR) are used for tracing changes of DNA and/or RNA differing in type of compound and mechanism used for detection (intercalating dyes, primers and probes). Different types of PCR were developed based on specific part of DNA being investigated. Besides PCR methods, several novel methods are employed. Next generation sequencing is used for obtaining information about localization and sequence of insert and its flaking regions. Microarray technology allows for multiple DNAs to be analyzed simultaneously on the chip. Decision-support system that detect and quantify GMO based on the Ct and Tm values and the LOD and LOQ is used in so-called matrix based methods. In so-called "food forensics" parameters like protected designation of origin (PDO) and protected geographic indication (PGI) that serve as an indication of food quality are being determined. In PGI determination several methods can be employed depending from type of molecule or parameters being analyzed, e.g. mass spectrometry for determining isotope ratio, spectroscopy or chromatography

How to cite this book chapter:

Pavlićević, M. and Vucelić-Radović, B. 2019. Application of PCR in Food Biochemistry. In: Vucelić Radović, B., Lazić, D. and Nikšić, M. (eds.) *Application of Molecular Methods and Raman Microscopy/Spectroscopy in Agricultural Sciences and Food Technology*, Pp. 97–117. London: Ubiquity Press. DOI: https://doi.org/10.5334/bbj.g.

for monitoring changes in lipid profiles etc. For determination of PDO, PCR methods that trace unique sequences, like single nucleotide polymorphisms (SNPs), restriction fragment length polymorphisms (RFLPs), amplified fragment length polymorphisms (AFLP), simple sequence length polymorphisms (SSLPs) are used. Novel method called DNA barcoding uses markers (short sequence complementary to target region) that can identify variation among cultivars. Today, in food forensics genomics, proteomics and metabolomics for determining authenticity are used. Both for DNA and RNA extraction, due to "matrix effect" choice of extraction method is of crucial importance for obtaining high amount of non-degraded pure nucleic acid. Thus, although both extraction of DNA and RNA consist of same basic steps (homogenization, lysis and extraction, purification (precipitation or binding), elution or resolubilization), lysis and extraction step will have the most influence on quality of isolated nucleic acid. During DNA extraction it is possible to use either kits (that are developed for specific purpose) or "traditional methods" were organic solvents are used for extraction. Today, for extraction of RNA mostly kits developed for RNA extraction from specific organisms and/or tissue are used. Since RNA is chemically more unstable than DNA it can easily be degraded during sampling and homogenization. For monitoring gene expression, Real Time PCR is used. Design of primer is possible in on-line free software such as Primer3 and Primer3Plus from NCBI or in software that are supplied with Real time PCR apparatus. Analysis of primers (possibility of formation of secondary structure, like hairpins and/or primer-dimmers) can be done in software like Vector NTI and MP primer.

1 "Food forensics"

Since in recent years safety and quality of food has become hot topic, determination of protected designation of origin (PDO) and protected geographic indication (PGI) as parameters for food quality has become imperative.

In determination of geographical origin of food, several methodologies classified by type of molecule being analyzed and parameters that are monitored have been employed (Luykx et al. 2008). In mass spectrometry methods composition of sample is determined by ionization of sample and subsequent measurement of ions m/z ratio and production of mass spectrum. So-called isotope ratio mass spectrometry has successfully being employed in authentication of food of animal origin by examining ratio of isotopes (Vinci et al. 2013). Isotopes ratio depends of several factors, such as animal diet, use of fertilizers etc., but among them geographical factors (such as soil composition, altitude etc.) play crucial role (Luykx et al. 2008). Spectroscopic methods such as infrared spectroscopy, can be used either to monitor production process like ripening in cheeses (Woodcock et al. 2008) or for determining authenticity by measuring changes in, for example, fatty acids profile, sterol and/or phenolic composition, volatile compounds, etc. Chromatographic methods, such as gas chromatography or high pressure liquid chromatography, can be used for determining authenticity of plant oils by, for example, determining its diacylglycerol and triacylglycerol profiles (Cserháti et al. 2008).

Food forensics employs genomics, proteomics and metabolomics for determining authenticity (Primrose et al. 2010). Mayor advantage of this integrative approach is that results of each individual method are verified in subsequent steps and limitations of particular technique are overcome by employing several levels of analysis.

PCR methods that are used for food authentication are based on tracing unique sequences (so-called DNA fingerprinting). Such methods include single nucleotide polymorphisms (SNPs), restriction fragment length polymorphisms (RFLPs), amplified fragment length polymorphisms (AFLP), simple sequence length polymorphisms (SSLPs) and, the use of real-time PCR and heteroduplex analysis (Primrose *et al.* 2010, Agrimonti *et al.* 2011). All of these methods detect small variation in base-pair sequence ("polymorphism").

In new approach, called DNA barcoding, markers (short sequence complementary to target region) are used to indentify not only species, but variation among cultivars (Galimberti et al. 2013). Although DNA barcoding is rapid, cheap and sensitive method, applying this technique on highly processed food can be unreliable, due to loss of DNA integrity (Galimberti et al. 2013).

Nowdays, for determination of protected designation of origin, genomic methods are combined with hysic-chemical detection. For example, Ganopoulos et al. (2011) used detection of microsatellite sequences with subsequent analysis by capillary electrophoresis and high resolution melting in determination of protected designation of origin of sweet cherry products.

2 Detection of GMO in food

Detection of genetically modified organisms in food could be achieved by employing several methods. Generally, these methods can be divided based on in what type of molecule changes is being detected. For tracing changes on protein level, enzyme-linked immunosorbent assay (ELISA), lateral flow sticks, Western blot and 2D electrophoresis are used. For tracing changes on DNA and/or RNA level, PCR methods are used. Broadly, PCR methods can be divided depending on their application and mechanism which is used for detection.

In so-called quantitative PCR (qPCR) that is used for screening of GMO, three types of chemistries are used. Intercalating dyes are molecules which aromatic ring(s) which are able to insert themselves between bases in DNA (usually in major groove). PCR methods using intercalating dyes (such as SYBR Green) are the cheapest, but sensitivity with single stranded DNA is low and in higher concentration they inhibit PCR reaction (Gasparic et al., 2010). There are several different chemistries currently employed in primer-based PCR. Their basic principle is the same: fluorescent dye (fluorofore) is attached to one end of primer and fluorescent signal is changed (decreased or increased) following annealing and extension step. However, crucial differences in these methods are in additional presence of synthetic bases (like in Plexor primers) or quenchers on the other end of primers (like in AmpiFluor primers). Such modifications enhance sensitivity and specificity of PCR reactions. Although, primer-based chemistry is more expensive than chemistry employing intercalating dyes, it is more sensitive and with higher specificity compared to SYBR Green PCR and it also gives possibility of multiplex PCR (when 4 different fluorofores, each specific to particular base, are used). However, primer-dimers and unspesific amplicons will also be detected (Gasparic et al. 2010). In probe-based chemistries (such as TaqMan) additional nucleotide sequence complementary to target sequence between primers (called probe) is used. Probe is marked with quencher at one end and fluorofore at the other end. After the extension step, quencher and fluorofore are realeased and absorbance is increased. TaqMan probe PCR is more expensive than SYBR Green PCR and primer-based PCR, but it has higher dynamic range, amplification efficiency and repeatability (Gasparic *et al.* 2010). Minor groove binding TaqMan probe is a variation of TaqMan PCR where TaqMan probe is bound to minor groove of DNA which in turn makes DNA-probe complex more stable and enhances specificity. Additional probe designs such as locked nucleic acid probe (LNA), cycling probe technology (CPT), and molecular beacons are also used.

The gene-specific PCR is more rigorous and more sensitive approach compared to real time PCR (Wen-Tao *et al.* 2009). It allows for multiplication of specific gene that codes for protein that can alter properties of particular cultivar or food product. For example, p35 gene in tomato codes for inhibitor of caspases, inhibiting cell death and providing for protection against diseases (Lincoln et al, 2002).

Target for construct-specific PCR is junction between two elements in DNA, usually between promotor and transgene. Therefore, positive signal will be present only in GMO material. However, if the same "foreign DNA" is present in different samples, difference in samples can't be detected (Wen-Tao et al. 2009). This type of PCR can be used, for example, for detection of *Bt* gene (coding for cytotoxin *Cry A*) in Roundup crops (Mesnage et al. 2013).

Since event-specific PCR targets locus between recipient and inserted DNA, it is even more specific and robust than construct-specific PCR (Randhawa et al. 2016). Event-specific PCR is often used for detection of Roundup Ready soybean and Bt-176 "Maximizer" maize in food (Berdal et al. 2001) even in highly-processed foodstuffs (Tengel et al. 2001).

To obtain information about localization and sequence of insert and its flaking regions, next generation sequencing is employed. Mayor difference between "classical" sequencing (by capillary electrophoresis) and next generation sequencing is that instead sequencing single DNA fragment in next

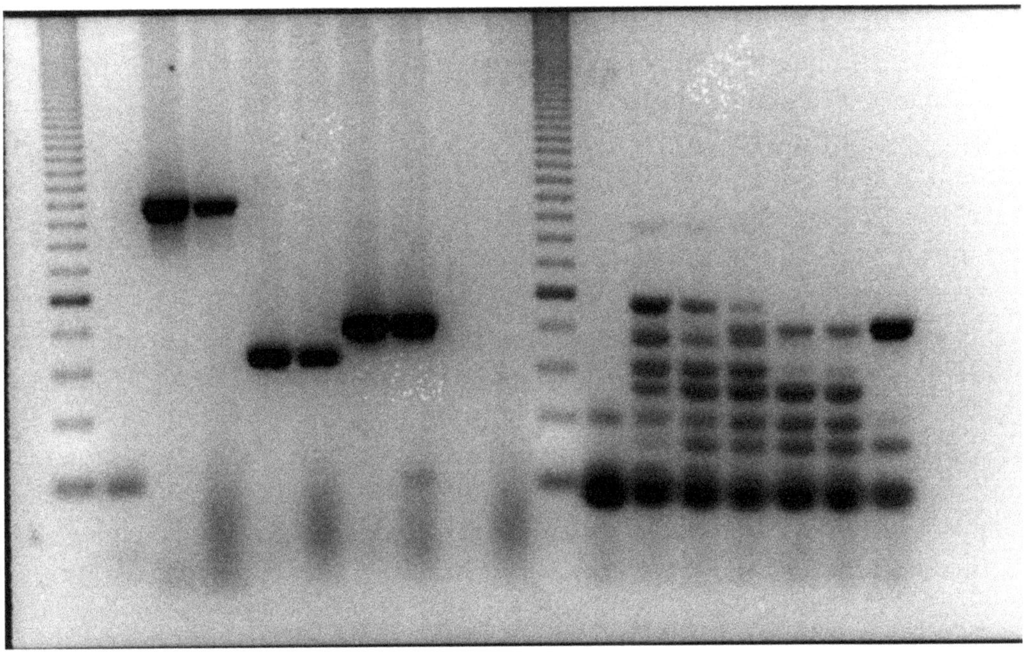

Figure 1: From left to right: lanes 1 and 11. 50-bp molecular mass markers; 2. PCR blank; 3 and 4. PCR products obtained by amplifying, in the presence of primer pairs (441 bp) for detecting DNA from Roundup Ready soybean; 5 and 6. Roundup Ready soybean (156 bp); 7 and 8. soybean lectin (210 bp); lanes 12–18. multiplex PCR products obtained by amplifying, in the presence of seven primer pairs for the simultaneous screening of endogenous genes of soybean and maize and five recombinant DNA constructs of genetically modified crops: Roundup Ready soybean and Bt11, Bt176, GA21, and MON810 lines of maize.

generation sequencing a large number of DNA fragments are sequenced simultaneously. Next generation sequencing coupled with site-finding PCR allowed detection of vip3Aa20 gene (coding for insecticidal protein) in MIR162 maize (Randhawa et al. 2016).

In microarray technology, DNA is attached to surface in form of microscopic spots. The biggest advantages of microarray technology are high sensitivity and high throughout (Berdal et al. 2001). Since different sequences can be analyzed simultaneously, there are different variations of microarray methodology (multiplex microarray, ligation detection reaction coupled with universal array technology, event-specific microarray, etc). For example, an event-specific DNA microarray was used to detect multiple GM events in processed food (soybean, maize, canola and cotton) (Kim et al. 2010).

Matrix based methods, like "Combinatory qPCR SYBR Green screening", use decision-support system that detect and quantify GMO based on 4 values: the Ct and Tm values and the LOD and LOQ (Van der Bulcke *et al.* 2010). Ct or "threshold cycle" represents a level above which fluorescence of sample exceeds background fluorescence. Ct value is inversely proportional to amount of DNA. Tm is a melting temperature. LOD represents limit of detection, while LOQ represents limit of quantification.

In contrast to previously mention methods, loop-mediated isothermal amplification (LAMP) is performed at constant temperature. Characteristic of LAMP is that 4 different primers are used for detection of 6 distinct regions on the target gene, with additional loop primers that accelerate reaction. Presence of multiple primers enhances sensitivity. During last few years, real-time LAMP assays for detection of two major commercialized Bt cotton events, MON531 and MON15985 were developed (Randhawa et al. 2016).

Key factor determining not only method for detection of GMO, but also for extraction is quality of DNA. During technological processing, quality, quantity and purity of DNA are all affected. For example, at lower pH (pH 3–4), glycosilic bonds between base and ribose are broken (Gryson et al. 2010). At pH above pH 8, the tautomeric state of the bases is affected. These tautomers can form non-standard base pairs that fit into the double helix and can cause the introduction of mutations during DNA replication. High temperature causes depurination and deamination which lead to degradation of DNA (Gryson et al. 2010). The effect of both temperature and water pressure during autoclaving has stronger degrading effect than cooking. For example, after autoclaving of soybeans at 121°C for 15 min only DNA fragments shorter than 295 bp are obtained (Ogasawara et al. 2003). Baking experiments at different temperature showed that baking reduces the size of the extracted DNA (Hrncirova et al. 2008).

2.1 Extraction of DNA from food

2.1.1 Introduction:

Nowadays, there are a variety of different methods for extraction of DNA from food sources that are being employed. In order to account for so-called "matrix effect" choice of extraction method largely depends on type ofsample that is being analyzed and degree of processing. However, goal of each of these methods is to gain high amount of non-degraded pure DNA of sufficient length. These requirements for intact DNA of substantial size stem from the fact that isolated DNA usually needs to be amplified for further experiments.

Although different in types of chemicals that are used and the mechanism by which DNA is isolated, each protocol for extraction of DNA from food samples consists of several basic steps:

1. Homogenization
2. Lysis and extraction
3. Purification (precipitation or binding)
4. Elution or resolubilization

Homogenization step serves for two purposes: to disrupt cells membranes (and/or cells walls) and to obtain the homogeneous, representative sample. Although different methods of homogenization exist (mortar and pestle, bead mills, rotor-stator homogenizator etc.) all of them use mechanical force. The impact of homogenization method on quality of obtained DNA is scarcely known, but several researches suggest (Colatat, Miller et al. 1999) that method of homogenization affects yield and purity as well as the fragment size of isolated DNA. In terms of yield and purity of DNA, homogenization with bead mill was found to be the best method, but rotor-stator homogenization gave the best results when it comes to fragment length. Food samples that contain large amount of compounds that can act as PCR inhibitors (lipids, phenolic compounds, polysaccharides etc. (Wilson et al. 1997) require pre-treatment (Terry et al. 2002). Methods for elimination of such contaminants depend on contaminant type (Terry et al. 2002). For example, lipids could be removed by hexane.

In lysis and extraction step cell wall or/and cell membrane that were disrupted in previous step are now being dissolved and cell releases its content, thus facilitating extraction of DNA from nucleus. For this purpose, lysis buffers are used and although their composition might vary depending on extraction method, they must contain following components: detergents (compounds able to remove/solubilize membrane lipids), RNAses (enzymes that degrade RNA) and proteinases that denaturize proteins. Often lysis buffer contains salt (for disturbing ionic bonds between proteins, mainly histones, and DNA) and chelating agent (that prevents degradation of DNA by binding ions like Mg^{2+} that serve as cofactors for DNAses). Generally, new methods for DNA extraction are mainly based on use of so-called kits. Two major types of kits used for extraction of DNA from food samples are: silica-based kits (like Wizard, NucleoSpin Food, QIAamp DNA Stool (Turci et al. 2010, Pirondini et al. 2010, Tung Nguyen et al. 2009) and kits that contain magnetic beads (like PrepFiler™ Forensic DNA Extraction Kit). In "traditional methods" organic solvents (e.g. phenol, chloroform, ethanol etc.) are used for extraction. Of these methods, SDS (sodium dodecyl sulfate) extraction and CTAB (cetyl trimethylammonium bromide) extraction are still used and modified today. Both of these methods give a high amount of DNA (Turci *et al.* 2010) and use relatively cheap chemicals, which are the main reasons behind their popularity. However, by comparing SDS and CTAB extraction, it could be concluded that CTAB method gives high yield even with thermally or chemically processed samples (such as cheese or cooking cream) while the same is not true for SDS extraction (Turci et al. 2010, Pirondini et al. 2010, Tung Nguyen et al. 2009). Also, with such samples that contain a high number of PCR inhibitors, SDS method gives a fragmented DNA (Turci et al. 2010, Pirondini et al. 2010). However, since modification of original CTAB procedure often involve transferring of sample from one tube to another, that can cause loss of the sample and increase risk of contamination. Also, both SDS and CTAB methods are time-consuming. Since CTAB method employs use of aggressive chemicals it gives DNA of lower quality compared with some novel methods such as Wizard (Turci *et al.* 2010). Mayor advantage of new methods is that they are fast. However, since most of the kits were developed for specific purpose, they don't give DNA of substantial yield or good quality with samples with higher degree of processing or with samples that greatly differ in their chemical composition from samples for which kit was developed. For example, with fresh and processed tomato products, NucleoSpin Food kit gave highest yield of all examined kits, but DNA was more degraded when compared to QIAamp DNA Stool and Wizard kits. Additional problem with usage of kits for DNA extraction is a high cost of chemicals that are needed for binding and purification of DNA.

During purification, DNA is separated from contaminants in solution. With traditional methods, this is done by precipitation of DNA. However, these methods differ in their ability to remove particular types of contaminants. Relatively high concentration of salt in CTAB buffer prevents co-precipitation of polysaccharides with DNA. However, CTAB is not as efficient as SDS in denaturating proteins and therefore proteins can remain as contaminants in precipitate. Both silica-based kits and kits with magnetic beads use negative charge of DNA for purification. Mechanism of purification of DNA by silica-based kits is not yet fully understood, but it is considered that chaotropic salts present in buffer are involved in formation of salt bridges between silica membrane and DNA. In complex mixtures such as food this can present a problem, since different compounds with same charge as DNA might remain in solution. Therefore, they will bind with DNA to the column and co-elute. Kits with magnetic beads are more successful in binding DNA, since negatively charged DNA is adsorbed on positively charged magnetic beads.

Since change in pH and/or ionic force is usually necessary for elution from membranes and columns, this step is also very sensitive, because such changes can fragment DNA. With traditional methods, resolubilization of precipitate might be an issue because usage of organic solvents might cause changes in DNA structure that can interfere with resolubilization of precipitate.

Since economics also plays significant part in choosing preferred method, a new approach called "fuzzy logic" is introduced in evaluation of methods for extracting DNA from food samples. That approach takes into account complexity of individual sample and ranks method not only by yield and quality of isolated DNA, but also by its cost.

Purity and concentration of isolated DNA can both be assessed using spectrometry to measure A_{260}/A_{280} ratio. Absorbance at 260 nm steams from aromatic rings in nitric bases in DNA, while absorbance at 280 nm is caused by presence of aromatic amino acids. Thus, this ratio is indicator of protein contamination. DNA with value higher than 1.8 for A_{260}/A_{280} ratio is generally considered as pure.

Quality of isolated DNA is determined by agarose electrophoresis and/or number of successful PCR reactions.

2.1.2 Materials, Methods and Notes

A. DNA extraction from milk using CTAB method (according to Corbispier et al. (2007))

Reagents needed:

CTAB extraction buffer (20 g of cetyl trimethylammonium bromide was dissolved in 1 l of 0.1 M Tris-HCl buffer pH 8 containing 1.4 M NaCl and 20 mM EDTA)

CTAB precipitation buffer (5g of trimethylammonium bromide was dissolved in 1 l of 40 mM NaCl)

RNAse A solution (100 mg/mL)

Proteinase K solution (20 mg/ml)

1.2 M NaCl

Chloroform

Ethanol (absolute and 70 % (v/v))

Nuclease free water

Procedure:

1. Incubate 100 μl of sample with 300 μl of nuclease-free water, 700 μl of CTAB extraction buffer and 5 μl of RNase A solution at 65° C for 15 min.
2. Add 20 μl of proteinase K solution and incubate for 15 min at 65°C.
3. Centrifuge for 10 min at 12000 g.
4. Transffer supernatant to a new tube and mix with 500 μl of chloroform
5. Transffer aqueous phase (700 μl) to new tube and mix with 700 μl of chloroform
6. Centrifuged for 5 min at 12000 g
7. Transffer suppernatat in new tube and incubate with with a double volume of CTAB precipitation buffer for 1 h at room temperature
8. Centrifuge for 15 min at 12000 g
9. Ressuspend pellet (DNA) in 400 μl of 1.2 M NaCl.
10. Resuspended DNA was mixed with 400 μl of chloroform
11. Centrifuge for 10 min at 12000 g
12. Transffer aqueous phase in new tube and mix with a double volume of ice-cold absolute ethanol.
13. Incubate for 20 min at 20°C and then centrifuged for 15 min at 12000 g.
14. Wash pellet (2x) with 500 μl of 70 % ethanol and air-dry
15. Dissolve in 100 μl of nuclease-free water.

B. DNA extraction from milk using PrepFiler™ Forensic DNA Extraction Kit (according to manufacturer user guide, Applied Biosystem (2008))

Reagents needed:

PrepFiler™ Lysis Buffer
1 M DTT (dithiothreitol)
PrepFiler™ Magnetic Particles
Isopropanol
PrepFiler™ Wash Buffer
PrepFiler™ Elution Buffer

Procedure:

1. 40 μl of sample was mixed with 300 μL of PrepFiler™ Lysis Buffer and 3 μLof 1M DTT in 1.5 ml tube and vortex for 5 s
2. Incubate for 20 min at 70°C at 900 rpm.
3. Centrifuge for 2 s at 14000 rpm.
4. Using pippette, transffer solution into PrepFiler™ Filter Column that is placed into 1.5 mL PrepFiler™ Spin Tube
5. Centrifuge for 2 min at 14000 rpm.
6. Prepare magnetic particles by vortexing the PrepFiler™ magnetic particles tube for 5 seconds at low speed and centrifuging for 2 s at 14000 rpm.
7. Disscard filter column and Pipette 15 μL of magnetic particles into 1.5 mL PrepFiler™ Spin Tube.
8. Vortex spin tube at 1200 rpm for 10 seconds and centrifuge for 2 s at 14000 rpm.
9. Add 180 μL of isopropanol and vortex for 5s at low speed .
10. Centrifuge for 2 s at 14000 rpm.
11. Mix at room temperature at 1000 rpm for 10 minutes.
12. Vortex for 10 s at 14000 rpm.
13. Centrifuge for 2 s at 14000 rpm.
14. Place the tube in the magnetic stand and wait for 1–2 min until size of the pellet remain constant
15. Discard supernatant carfully (not to disturb DNA in pellet)
16. Wash the pellet 3x by pipetting 300 μL of PrepFiler™ Wash Buffer, vortexing 5 s at maximum speed and centrifuging for 2s at 14000 g. Supernatant is discard after each washing step
17. Place sample tube in the magnetic stand, open it and allow DNA to air-dry for 7 to 10 minutes at room temperature.
18. Add 50 μL of PrepFiler™ Elution Buffer and vortex for 5 s at maximum speed
19. Centrifuge for 2 s at 14000 rpm.
20. Incubate at 70°C at 900 rpm for 5 minutes.
21. Vortex 5 s at maximum speed.
22. Centrifuge for 2 s at 14000 rpm.
23. Place the tube in the magnetic stand and wait until the size of the pellet stops increasing (at least 1 minute).
24. Pippete suppernatant (that contains sample DNA) into new 1.5 ml tube.

25. *Additional step*: if the eluted DNA extract is turbid, centruge for 5 min at 14000 rpm and transffer supernatant into new tube.

2.2 Extraction of RNA from food

2.2.1 Introduction:

Basic steps in RNA extraction are the same as for DNA (homogenization, lysis, extraction and purification and elution). However, RNA is chemically more unstable compared to DNA, thus it is prone to degradation by RNAses (Lepinske et al. 1997) and sensitive to changes in temperature. Therefore, when working with RNA, it is crucial to prepare and maintain RNAses-free environment, by inhibiting enzymes with alcohol and low temperatures. Unlike DNAses, RNAses don't need addition of divalent ions for their activity. Further inhibition of endogenous RNAses is achieved by adding either 2-mercaptoethanol and/or guanidinium salts in extraction (lysis) buffer. 2-mercaptoethanol inhibits enzymatic action by breaking disulfide bonds in enzymes. Guanidinium salts also denature proteins, causing inhibition of RNAses.

Because most of the potential degradation of RNA by RNAses happens during sampling and homogenization, these two steps are crucial for obtaining non-degraded RNA of sufficient quantity (MacRae 2007). During sampling care must be taken to quickly place sample in liquid nitrogen in order to avoid degradation. Although it was thought that because of higher amount of RNA younger plant tissues are better choice, recent experiment (Johnson et al. 2012) showed that difference in amount of total RNA among older and younger tissues was only around 13%. Also, not all plant tissues are equally metabolically active and thus don't contain same amount of RNA and/or same amount of inhibitors (such as phenolic compounds, among them especially tannins, polysaccharides, proteins, etc). During homogenization, there are a few possible obstacles in extracting RNA. Firstly, short time and/or not enough force during disruption will not release all RNA from cells in samples, thus leading to lower yield. Secondly, environment that is not exogenous RNAses-free will cause degradation of RNA during homogenization. Thirdly, during this step loss of sample might occur.

Extraction of RNA might be done either by commercial kits or by extraction with organic solvents. Of the latter, very popular is so-called Trizol extraction that uses guandinum thiocyanate, phenol and chloroform for extraction and purification of RNA. However, as every method it has its setbacks such as not giving good results with plant tissues that are rich in polysaccharides and phenolic compounds and it might be hard to dissolve the pellet (Rio *et al.* 2010). In most cases, extraction procedures are adjusted for a specific tissue or a purpose. Most of these modifications consider removal of specific inhibitors that would otherwise co-precipitate with RNA or bind with it. Among these inhibitors, polysaccharides and phenolic compounds are the most prominent. Polysaccharides can cause formation of slurry during resuspension (Dal Cin et al. 2005). Unlike with DNA extraction, enzymes like pectinase could not be used during extraction of RNA because of the risk of degradation (Dal Cin et al. 2005). Some protocols are adjusted not for a type of tissue but for a purpose. For example, in gene expression studies, it is necessary to modify a method to be quick, to work well with tissues with varying levels of inhibitors and to avoid steps that can affect yield (Vasanthaiah et al. 2008). Similar to DNA extraction, basic principle of RNA extraction is binding to silica-based column or membrane. Such binding can be based on negative charge of RNA or on differences in molecular masses between RNA and inhibitors in the extract.

During extraction of RNA by organic solvents, precipitation of pellet containing RNA is necessary. Resuspension of pellet could be a problematic step because it is known that presence of secondary metabolites could lower solubility of pellet after extraction involving guanidinium salts (Ghawana et al. 2011).

2.2.2 Materials, Methods and Notes

A. Extraction of total RNA from leaves and roots of *Arabidopsis thaliana* (according to RNeasy Mini Handbook Qiagen)

Reagents needed:
Buffer RLT
Buffer RW1
Buffer RPE
RNase-free water

Protocol

1. Homogenize 100 mg of plant tissue (leaves or roots) in liquid nitrogen using mortar and pestle
2. Place grinded sample into 2 ml tube
3. Add 450 μl Buffer RLT (lysis buffer) and vortex at maximum speed
4. Transfer the lysate to a QIAshredder spin column that is already placed in a 2 ml tube and centrifuge for 2 min at maximum speed.
5. Transfer the supernatant into new 2 ml tube, without disturbing pellet
6. Add 500 μl of absolute ethanol and mix by turning the tube
7. Transfer entire sample into the RNeasy spin column that is already placed in a 2 ml tube. Close the lid.
8. Centrifuge for 15 s at 14000 rpm
9. After discarding flow-through, without changing tube, add 700 μl Buffer RW1 to the RNeasy spin column. Close the lid.
10. Centrifuge for 15 s at 14000 rpm

11. After discarding flow-through, without changing tube, add 700 µl Buffer RPE to the RNeasy spin column. Close the lid.
12. Centrifuge for 15 s at 14000 rpm
13. After discarding flow-through, without changing tube, add 700 µl Buffer RPE to the RNeasy spin column. Close the lid.
14. Centrifuge for 2 min at 14000 rpm
15. Place the RNeasy spin column in a new 1.5 ml collection tube.
16. Add 30–50 µl RNase-free water directly to the spin column membrane.
17. Close the lid and centrifuge for 1 min at 14000 rpm. RNA remains in solution.

Notes: If the gene expression is going to be tested, two additional steps (elimination of genomic DNA and construction of cDNA) are necessary.

The QuantiTect Reverse Transcription kit combines these steps.

B. cDNA synthesis using QuantiTect Reverse Transcription kit (according to QuantiTect Reverse Transcription Handbook)

Reagents needed:

gDNA Wipeout buffer
RNAse- free water
Quantiscript reverse transcriptase
Quantiscript RT buffer
RT primer mix

Protocol:

1. While keeping tube with sample DNA on ice, add 2 µl of wipeout buffer and with RNAse- free water adjuct volume, so that the total reaction volume is 14 µl
2. Incubate for 2 min at 42°C (during this time, genomic DNA eventualy present as contaminant is eliminated)
3. Place tube on ice and add 1 µl of Quantiscript reverse transcriptase, 4 µl of Quantiscript RT buffer and 1 µl of RT primer mix.
4. Incubate for 15 min at 42° C (during this step, cDNA is constructed)
5. Incubate for 3 min at 95°C (in order to denaturate reverse transcriptase).

3 Real time PCR

3.1 Introduction

At the moment real-time PCR is the only method that allows absolute quantification of nucleic acid. Its main application include: detection and quantification of pathogens (Kubista *et al.* 2006, Yang *et al.* 2004), detection of allergens in food (Parfundo et al. 2009), measurement of gene expression (Wong et al.

2005), etc. Depending of purpose, it is possible to perform real time PCR from DNA and/or from RNA as a template.

Regardless whether starting material is RNA or DNA, the most critical steps in every real time procedure are obtaining good quality sample and choosing "good" primers. Good quality sample is high amount of purified, non-degraded nucleic acid. In case when different expression of gene is monitored, additional care must be taken in removing all traces of genomic DNA from mRNA. Otherwise, during synthesis of cDNA and further amplification, falsely high results might be expected. Of course, since as for "conventional" PCR integral part is amplification of target sequence, all inhibitors (lipids, proteins, carbohydrates, etc.) that affect success of "conventional" PCR will also influence reaction of real-time PCR.

Beside specificity, primers for real-time PCR must be stable (characterized by Tm values and GC content) and of certain size (in order to avoid miss-paring or loss in amplificability). Since there are different types of real-time PCR (such as duplex, multiplex, etc.) for each of these techniques there are additional conditions for primers (Shen *et al.* 2010). However, irrelevant of type of real-time PCR, analysis of sequence of forward and reverse primers must be done in order to avoid creation of secondary structure within primer itself or creation of so-called primer-dimmers. Primer-dimmers are created when hybridization happen not between primer and template but between forward or reverse primers with each other. There are several software (e.g. Primer 3, MPrimer, etc.) which give possibility to choose the best primer sequence based on the sequence of target gene. In such software characteristics of primers, such as melting temperature, GC content and possibility of production of secondary structures (including primer-dimmers) are given. Specificity of primers and eventual existence of primer-dimmers can be examined by melting curve analysis. In ideal situation, only one peak should be present for each curve. Also, location of peak for each curve must be the same, e.g. curves should overleap. As a control some of genes which have a constant expression (because of its involvement of basic metabolic pathways) are usually used. Such genes are called housekeeping genes. Also, to check efficiency of amplification, internal amplification control is used. Internal amplification control represents non-target DNA fragment (of known concentration) that is added to the reaction and amplified in parallel with the target sequence.

There are two ways to monitor amplification process in real time. First approach uses SYBR Green chemistry. SYBR Green is fluorescent dye that intercalates with double stranded nucleic acid. Major disadvantage of SYBR Green approach is that it can bind to any double stranded nucleic acid and not just only to hybridization products between primer and target. Therefore, it will also bind to primer-dimmers. This binding decreases specificity of this method in comparison to TaqMan probe chemistry. TaqMan probe is a sequence-specific oligonucleotide probe that binds to a target in between forward and reverse primers. TaqMan probe has at each end attached a fluorescent dye-at

5' end –fluorofore and at 3' end-quencher. Before PCR cycle fluorescence of fluorofore is "quenched" (inhibited) by quencher. During cycle, as DNA polymerase moves from 3' to 5' end, it encounters and hydrolyzes fluorofore from 5' end. Since emission of unbound fluorofore is no longer inhibited, fluorescence starts to increase. Sequence of TaqMan probe might be determined by same software in which it is possible to determine sequence of primers. However, due to chemistry of reaction, conditions for "good" TaqMan probe are different from those for primers. To ensure stability and specificity, Tm value and length for TaqMan probe are usually higher for those of primers. Since using guanine near the 5' end of the TaqMan probe can quench fluorescence, it should be avoided. As for primers, sequence with consecutive stretches of the same base must be excluded (because it can affect hybridization efficiency due to formation of secondary structure).

Quantification is done by measuring fluorescence during cycles. Fluorescence is higher for higher concentration of product. Firstly, number of copies of DNA is insufficient for fluorescence to surpass threshold level. After few cycles, concentration of sample is sufficient for surpassing threshold level and fluorescence starts to increase exponentially. Because reaction efficiency is stable during exponential phase, a threshold level should be chosen so that it reflects data during this phase. After subtracting background from raw data, comparison of concentration of samples can be done.

3.2 Materials, Metodes and Notes

Procedure for Real time PCR (according to Getting Started Guide Applied Biosystems 7500/7500 Fast Real-Time PCR System Standard Curve Experiments)

A Preparation for experiment:

Two steps are necessary before measuring flourescence of sample and those are, of course, design of primers and prepartion of mix (layering the plate).

B Primer design:

Among on-line free based software for designing and analyzing samples, one of the most popular are Primer3 (Thornton et al. 2011) and Primer3Plus from NCBI (National Center for Biotechnology Information). However, designing and analyzing primers is also possible in software that are supplied with Real time PCR apparatus (with 7500/7500 Fast Real-Time PCR System, Primer Express v.3 software is supplied). Although procedure for preparing design for Real Time PCR using TaqMan probe involve additional step(s) for probe design, there are few parametars that are neccessary to adjust to get succesfull

Real time PCR reaction. Those parametars are: primer lenght, melting temperature (T_m), GC content, product size, 3' stability.

Length of the primer: if the primer is to short, due to the lack of specificity, it will not hybridize with template, while if it is too long hybridization will take too long and it will be hard to remove product. Optimal primer length is 18–24 base pairs (bp). Difference in primer pair length should be less than 3bp.

Melting temperature (T_m) is defined as temperature at which 50% of the primer is hybridized with its template. Again, too high T_m will lead to secondary annealing, while to low T_m will impede with hybridization. Optimal T_m range is 57–65°C. Difference in T_m of forward and reverse primer should be less than 5°C (ideally 1°C).

GC content is defined as the number of guanine and cytosine bases, expressed as a percentage of the total bases in primer. GC content is a measure of primer stability (stability is expressed through ΔG value) and should be 40–60%. In addition to this, number of guanine and cytosine within the last five bases from the 3' end of primers is called GC clamp. GC clamp value should be less than 3, because although G and C residues at 3' end will enhance specific binding to template, such sequence will give amplicons of bigger sizes.

Size of product- fluorescence during SYBR® Green detection will be more intense if product is longer. Ideal amplicon size is between 80 and 150 bp. Shorter amplicons are amplified more efficiently than longer ones.

3' stability is expressed as maximum ΔG value of the last five bases at 3' end of the primers. Higher 3' stability means improved efficiency of the primers. Also, some factors, as for example concentration of divalent ions, might need to be changed (usually SYBR® Green buffer mixes contain 3 to 6 mM of Mg^{2+}). During primer synthesis it is essential to check number of repeats and runs, since higher number of repeats and/or runs will lead to miss-priming or/and formation of primer secondary structures. Repeats are repeated sequences of dinucleotide (maximum number of repeats should be 4). Runs are defined as single nucleotide sequence repeats and their preferred maximum number is 3–4 bp. Also, when working with TaqMan probe, it is necessary to adjust T_m of the probe to around 10°C higher than T_m of primers (because the minor groove binding will increase T_m of the probe) and size should be less than 30 bp (not to affect specificity). Also, guanine residue at 5' end should be avoided because it can quench fluorescence.

Analysis of primers can be done in software like Vector NTI and MP primer. In this software there is an option to check for possibility of formation of secondary structure, like hairpins and/or primer-dimmers.

Steps

1. Finding and coping the sequence in FASTA Format NCBI website (http://www.ncbi.nlm.nih.gov/guide/) → nucleotide database (from dropdown menu) → in search box enter gene code → search → display settings → FASTA→ apply Copy FASTA sequence into new word document.

2. NCBI website → tools→ BLAST→ primer BLAST or alternatively go to Primer3 website (http://bioinfo.ut.ee/primer3–0.4.0/)
3. Copy sequence from word file in PCR template (or in sequence ID in Primer3)
4. Either use your own primers or give a range for forward and reverse primers (in primer 3 check boxes: Pick left primer, or use left primer below and Pick right primer, or use right primer below; if you are working with TaqMan also check box Pick hybridization probe)
5. In exon/intron selection, cut the intron part (In primer3 →excluded regions)
6. Change primer parameters or leave it default (In Primer3→ General primer picking conditions)
7. Click get (pick) primers

C Layering the plate

Microtitar plate with 96 wells is used. Beside samples, 2 internal standards and blanc probe should also be loaded in wells. Blanc (negative) probe contains all that is in sample, except cDNA. Internal standards are so-called housekeeping gens that code for proteins that are expressed in all cells, since they are involved in basic cellular processes (e.g. gen for GAPDH). In sample wells load: 25 μl 2x SYBER Green PCR Master Mix + 6.2 μl forward primers + 6.2 μl reverse primers + 6.2 μl cDNA sample + 6.4 μl RNAse-free H_2O. However, concentration of water, primers and sample is variable and should be optimized for particular reaction.

D Run

Cover the plate with transparent foil and centrifuge for few seconds (Centrifugation is done in order to avoid air bubbles).

Load the plate in plate holder and align.

By pushing, close the tray door.

In Setup page, choose run option.

In Run Method, it is possible to select number of cycles and change run parameters. Default run settings:

1. pre-PCR phase –60°C, 30s (elongation)
2. holding stage –95°C, 10 min (activation of Hot-Start DNA polymerase)
3. cycling stage –92°C,15s (melting of DNA); followed by 90s at 60°C (annealing of primers to templates and in the case of TaqMan Real time PCR, also the probe).

Click start run button.

Note: Run can be monitored by selecting amplification plot and choosing option of view plate layout.

4 Acknowledgments

Authors are grateful to AREA project (FP7-REGPOT-0212-2013 –I) and University of Parma, Department of Life Sciencies for providing training and expertise.

5 References

Agrimonti, C., Vietina, M. Pafundo, S., & Marmiroli, N. (2011).The use of food genomics to ensure the traceability of olive oil. *Trends in food science and technology*, 22, 237–244. DOI: https://doi.org/10.1016/j.tifs.2011.02.002

Berdal, K.G., & Holst-Jensen, A. (2001). Roundup Ready soybean event-specific real-time quantitative PCR assay and estimation of the practical detection and quantification limits in GMO analyses *European Food Research and Technology*, 213, 432–438. DOI: https://doi.org/10.1007/s002170100403

Colatat, M. Effect of Homogenization Method on DNA Yield and Fragment Size. Retrieved from http://opsdiagnostics.com/applications/nucleicacids/homogdnacompare.htm

Corbisier, P., Broothaerts, W., Gioria, S., Schimmel, H., Burns, M., & Baoutina, A. (2007). Toward metrological traceability for DNA fragment ratios in GM quantification. 1. Effect of DNA extraction methods on the quantitative determination of Bt176 corn by Real-Time PCR. *Journal of Agriculture and Food Chemistry*, 55, 3249–3257. DOI: https://doi.org/10.1021/jf062931l

Cserháti, T., Forgács, E., Deyl, Z., & Miksik, I. (2005). Chromatography in authenticity and traceability tests of vegetable oils and dairy products: a review. *Biomedical Chromatography,* 19, 183–190. DOI: https://doi.org/10.1002/bmc.486

Dal Cin, V., Danesin, V., Rizzini, M. F., & Ramina, A. (2005). RNA Extraction From Plant Tissues. The Use of Calcium to Precipitate Contaminating Pectic Sugars. *Molecular biotechnology,* 31: 113–119. DOI: https://doi.org/10.1385/MB:31:2:113

Galimberti, A., De Mattia, F., Losa, A., Bruni, I., Federici, S., Casiraghi, M., Martellos, S., & Labra, M. (2013). DNA barcoding as a new tool for food traceability. *Food research international*, 50, 55–63. DOI: https://doi.org/10.1016/j.foodres.2012.09.036

Ganopoulos, I., Argiriou, A., & Tsaftaris, A. (2011). Microsatellite high resolution melting (SSR-HRM) analysis for authenticity testing of protected designation of origin (PDO) sweet cherry products. *Food Control,* 22, 532–541. DOI: https://doi.org/10.1016/j.foodcont.2010.09.040

Gasparic, M. B., Tengs, T., La Paz, J. L., Holst-Jensen, A., Pla, M., Esteve, T., Zel, J., & Gruden, K. (2010). Comparison of nine different real-time PCR chemistries for qualitative and quantitative applications in GMO detection. *Analitical and Bioanaltical Chemistry*, 396, 2023–2029. DOI: https://doi.org/10.1007/s00216-009-3418-0

Getting Started Guide Applied Biosystems 7500/7500 Fast Real-Time PCR System Standard Curve Experiments (6), Part Number 4387779 Rev. C 06/2010. Retrieved from http://www3.appliedbiosystems.com/cms/groups/mcb_support/documents/generaldocuments/cms_050329.pdf

Ghawana, S., Paul, A., Kumar, H., Kumar., A., Singh, H., Bhardwaj, P.A., Rani, A., Singh, S.R., Raizada, J., Singh, K., & Kumar, S. (2011). An RNA isolation system for plant tissues rich in secondary metabolites. *BMC Research Notes*, 4, 85. DOI: https://doi.org/10.1186/1756-0500-4-85

Gryson, N. (2010). Effect of food processing on plant DNA degradation and PCR-based GMO analysis: a review. *Analitical and bioanalitical chemistry*, 396, 2003–2022. DOI: https://doi.org/10.1007/s00216-009-3343-2

Hrncirova, Z., Bergerova, E., & Siekel, P. (2008). Effects of technological treatment on DNA degradation in selected food matrices of plant origin. *Journal of Food and Nutrition Research*, 47(1), 23–28.

Johnson, T. J. M., Carpenter, J. E., Tian, Z., Bruskiewich, R., Burris, N.J., Carrigan, C., Chase, W.M., Neil, D.C., Covshoff, S., de Pamphilis W. C., Edger, P.P., Goh, F., Graham, S., Greiner, S., Hibberd, J. M., Jordon Thaden I., Kutchan, T. M., Leebens-Mack, J., Melkonian, M., Miles, N., Myburg, H., Pires, J.M., Ralph, P., Rolf, M., Soltis, D., Soltis, P., Stevenson, D., Stewart, C. N. J., Thomsen, J. M. C., Villarreal, J. C., Wu, X., Zhang, Y., Sage, R. F., Surek, B., Deyholos, M. K., & Ka-Shu Wong, G. (2012). Evaluating Methods for Isolating Total RNA and Predicting the Success of Sequencing Phylogenetically Diverse Plant Transcriptomes. *PLoS ONE* 7(11), e50226, DOI: https://doi.org/10.1371/journal.pone.0050226

Kim, J. H., Kim, S. Y., Lee, H., Kim, Y. R. & Kim, H. Y. (2010). An event-specific DNA microarray to identify genetically modified organisms in processed foods. *Journal of agricultural and food chemistry*, 58(10), 6018–6026. DOI: https://doi.org/10.1021/jf100351x

Kubista M., Andrade J. M., Bengtsson M., Forootan A., Jonák J., Lind K., Sindelka R., Sjöback R., Sjögreen B., Strömbom L., Ståhlberg A., & Zoric N (2006). The real-time polymerase chain reaction. *Molecular Aspects of Medicine*, 27, 95–125. DOI: https://doi.org/10.1016/j.mam.2005.12.007

Lepinske, M. (1997). Tips for Working with RNA and Troubleshooting Downstream Applications. *Promega Notes Magazine*, 63, 17–22.

Lincoln, J. E., Richael, C., Overduin, B., Smith, K., Bostock, R., & Gilchrist, D. G. (2002). Expression of the antiapoptotic baculovirus p35 gene in tomato blocks programmed cell death and provides broad-spectrum resistance to disease. *Proceedings of National Acadamy of Scienies* 99(23), 15217–15221. DOI: https://doi.org/10.1073/pnas.232579799

Luykx, M. A. M. D., & van Ruth, M. S. D. (2008). An overview of analytical methods for determining the geographical origin of food products. *Food Chemistry*, 107, 897–911. DOI: https://doi.org/10.1016/j.foodchem.2007.09.038

MacRae, E. (2007) Extraction of Plant RNA. In E. Hilario, & J. Mackay (Eds.), Methods in Molecular Biology (pp. 15–24). Protocols for Nucleic Acid Analysis by Nonradioactive Probes, 2nd Edition, Humana Press, Totowa.

Marmiroli, N. Peano, C., & Maestri, E. (2003). Advanced PCR techniques in identifying food components. In M Lees (Ed.), *Food authenticity and traceability* (pp. 1–33). CRC Press, Boca Raton Boston New York Washington, Woodhead Publishing Ltd. DOI: https://doi.org/10.1533/9781855737181.1.3

Mesnage, R., Clair, E., Gress, S., Then, C., Székács, A., & Séralini, G.E. (2013). Cytotoxicity on human cells of Cry1Ab and Cry1Ac Bt insecticidal toxins alone or with a glyphosate-based herbicide. *Journal of Applied Toxicology*, 33(7), 695–699. DOI: https://doi.org/10.1002/jat.2712

Miller, D.N., Bryant, J. E., Madsen, E.L., & Ghiorse, W. C (1999). Evaluation and optimization of DNA extraction and purification procedures for soil and sediment samples. *Applied and environmental microbiology*, 65(11), 4715–4724.

Ogasawara, T., Arakawa, F., Akiyama, H., Goda, Y., & Ozeki, Y. (2003). Fragmentation of DNAs of processed foods made from genetically modified soybeans. *Japanese Journal of Food Chemistry*, 10(3),155–160. DOI: https://doi.org/10.18891/jjfcs.10.3_155

Parfundo S., Gulli M., & Marmiroli N. (2009). SYBR-Green Real-time PCR to detect almond traces in processed food. *Food Chemistry*, 116, 811–815. DOI: https://doi.org/10.1016/j.foodchem.2009.03.040

Pirondini, A., Bonas, U., Maestri, E., Visioli, G., Marmiroli, M., & Marmiroli, N. (2010). Yield and amplificability of different DNA extraction procedures for traceabilityin the dairy food chain. *Food Control*, 21, 663–668. DOI: https://doi.org/10.1016/j.foodcont.2009.10.004

Primrose, S., Woolfe, M., & Rollinson, S. (2010). Food forensics: methods for determining the authenticity of foodstuffs. *Trends in food science and technology*, 21, 582–590. DOI: https://doi.org/10.1016/j.tifs.2010.09.006

QuatiTect Reverse Transcription Handbook, Qiagen, 03/2009

Randhawa, G. J., Singh, M., & Sood, P. (2016). DNA-based methods for detection of genetically modified events in food and supply chain. *Current science*, 110(6), 1000–1009. DOI: https://doi.org/10.18520/cs/v110/i6/1000-1009

Rio, D. C., Ares, M. Jr., Hannon, G. J., & Nilsen, T. W. (2010). Purification of RNA using TRIzol (TRI reagent). *Cold Spring Harbor Protocols* 6, RNeasy Mini Handbook, Qiagen, 06/2012. DOI: https://doi.org/10.1101/pdb.prot5439

Shen, Z., Qu, W., Wang, W., Lu, Y., Wu, Y., Li, Z., Hang, X., Zhao, D., Zhang, C. & Wang, X. (2010). MPprimer: a program for reliable multiplex PCR primer design. *BMC Bioinformatics*, 11, 143–149. DOI: https://doi.org/10.1186/1471-2105-11-143

Tengel, C., Schüßler, P., Setzke, E., Balles, J., & Sprenger-Haußels, M. (2001). PCR-based detection of genetically modified soybean and maize in raw and highly processed foodstuffs. *BioTechniques*, 31, 426–429. DOI: https://doi.org/10.2144/01312pf01

Terry, C., Harris, N. & Parkes, H. (2002). Detection of genetically modified crops and their derivatives: critical steps in sample preparation and extraction. *Journal of AOAC International*, 85(3), 768–774.

Thornton, B. & Basu, C. (2011). Real-time PCR (qPCR) primer design using free online software. *Biochemistry and molecular biology education*, 39(2), 145–154. DOI: https://doi.org/10.1002/bmb.20461

Tung Nguyen, C. T., Son, R., Raha, A. R., Lai, O. M., & Clemente Michael, W. V. L. (2009). Comparison of DNA extraction efficiencies using various methods for the detection of genetically modified organisms (GMOs). *International Food Research Journal*, 16, 21–30.

Turci, M., Sardaro, M., Visioli, G., Maestri, E., Marmiroli, M., & Marmiroli, N. (2010) Evaluation of DNA extraction procedures for traceability of various tomato products. *Food Control*, 21, 143–149. DOI: https://doi.org/10.1016/j.foodcont.2009.04.012.

User guide PrepFiler™ Forensic DNA Extraction Kit. Applied Biosystem, Part Number 4390932 Rev. B 11/2008, 2008. Retrieved from http://www3.appliedbiosystems.com/cms/groups/applied_markets_support/documents/generaldocuments/cms_053966.pdf

Van den Bulcke, M., Lievens, A., Barbau-Piednoir, E., MbongoloMbella, G., Roosens, N., Sneyers, M., & Casi, A.L. (2010). A theoretical introduction to "combinatory SYBRGreen qPCR screening", a matrix-based approach for the detection of materials derived from genetically modified plants. *Analitical and bioanalitical chemistry*, 396(6), 2113–23. DOI: https://doi.org/10.1007/s00216-009-3286-7

Vasanthaiah, K. N. H., Katam, R., & Sheikh, B. M. (2008). Efficient protocol for isolation of functional RNA from different grape tissue rich in polyphenols and polysaccharides for gene expression studies. *Electronic Journal of Biotechnology*, 11(3), DOI: https://doi.org/10.2225/vol11-issue3-fulltext-5

Vinci, G., Preti, R., Tieri, A., & Vieri, S. (2013). Authenticity and quality of animal origin food investigated by stable-isotope ratio analysis. *Journal of Science of Food and Agriculture*, 93(3),439–48. DOI: https://doi.org/10.1002/jsfa.5970

Wen-Tao, X., Wei-Bin, B., Yun-Bo, L., Yan-Fang, Y., & Kun-Lun, H. (2009). Research progress in techniques for detecting genetically modified organisms. *Chinese Journal of Agricultural Biotechnology*, 6(1), 1–9. DOI: https://doi.org/10.1017/S1479236209002575

Wilson, G. I. (1997). Inhibition and facilitation of nucleic acid amplification. *Applied and environmental microbiology*, 63(10), 3741–3751. DOI: https://doi.org/0099-2240/97/$04.0010

Wong, M. L. & Medrano, J. F. (2005). Real-time PCR for mRNA quantitation. *BioTechniques*, 39,75–85. DOI: https://doi.org/10.2144/05391RV01

Woodcock, T., Fagan, C., O'Donnell, C., & Downey, G. (2008). Application of near and mid-infrared spectroscopy to determine cheese quality and authenticity. *Food Bioprocess Technology*, 1, 117–129. DOI: https://doi.org/10.1007/s11947-007-0033-y

Yang, S., Rothman, R. E. (2004). PCR-based diagnostics for infectious diseases: uses, limitations, and future applications in acute-care settings. *Infectious Diseases*, 4, 337–348. DOI: https://doi.org/10.1016/S1473-3099(04)01044-8

Application of Molecular Methods in Aquaculture and Fishery

Zorka Dulić, Božidar Rašković, Saša Marić, Tone-Kari Knutsdatter Østbye

Abstract

Early aquaculture studies were mainly engaged in practising different culturing systems, primarily focusing on the improvement of fish growth and feed type. The significant advances in molecular biology during the last century greatly influenced the development of genetic research and application of molecular methods in aquaculture and fishery. These methods provided substantial opportunity for increased production efficiency, better product quality and improvement of animal health. Additionally, DNA based methods provided tremendous improvement in species determination, family tracking, pedigree determination of individuals, identification of lineages, identification of markers and loci that are responsible for economically important traits. In this work we pay special attention to the utilization of restriction fragment length polymorphism (RFLP) in aquaculture and fishery as a practical and relatively cheap method for determination of genetic lineages in polymorphic fish species. We also focus on gene expression as an excellent method for understanding physiological processes in fish. Currently, quantitative real-time PCR is one of the most accurate methods for gene expression analysis. It is very precise, sensitive

How to cite this book chapter:

Dulić, Z., Rašković, B., Marić, S. and Østbye, TK., K. 2019. Application of Molecular Methods in Aquaculture and Fishery. In: Vucelić Radović, B., Lazić, D. and Nikšić, M. (eds.) *Application of Molecular Methods and Raman Microscopy/ Spectroscopy in Agricultural Sciences and Food Technology*, Pp. 119–139. London: Ubiquity Press. DOI: https://doi.org/10.5334/bbj.h.

and gives relatively fast results. The growing knowledge in this area has enabled better understanding of fish physiology in aquaculture aiming to achieve economical feasibility and sustainability at the same time.

1 Introduction

Identification of fish species is traditionally based on external morphological characters as body shape, colour, scale size, position and size of fins as well as relative size of different body parts (Strauss & Bond 1990).

In species that are morphologically more similar number of gill rakers can be used as a significant systematic character. Furthermore, morphology of otoliths is also used in determination of fish age and size, but is especially useful for identification of the fish species from fossil remains or gut content of a predator (Granadeiro & Silva 2000).

However, in some cases the external morphology is not sufficient for the identification and determination even with whole specimens, either due to considerable intraspecific variations or small interspecific differences (Teletchea 2009). In addition, identification of fish species using morphology is not possible when specimens are semi degraded caused by digestion, when coming from the gut content or due to processing (e.g. canning or filleting). Identification of early life stages, as eggs or larvae, is more complicated that identification of adult stages. Due to all these difficulties, researchers were looking for new methods that would provide easier and more reliable identification of fish species.

Increased fish and seafood consumption worldwide caused Intensification of international trade and consequently varying levels of supply and demand of certain high quality fish or seafood species. Development of the fish market also led to economic fraud and the substitution of one species (usually more expensive) with another fish species or seafood product (less expensive). This was particularly successful when processed fish and seafood were used as fillets or canned products (Rasmunsen et al. 2008). Thus, it has become more and more important to know valuable facts about the products on the market and the authentication of fish and seafood species an important issue. In Europe, a law has been established for fish and aquaculture products requiring traceability information as identity and origin of fish as well as production method (Regulation (EC) No 104/2000).

Traditional and official methods for the identification of different species, including fish, have been based on the separation or characterization of specific proteins using techniques as electrophoresis, liquid and gas chromatography or utilization of immunological assays (ELISA) (Moretti et al. 2003, Kvasnicka 2005, Hubalkova et al. 2007). Although these methods are valuable in species identification, especially on fresh and frozen products, they are less reliable on products that are heat processed or dried, mainly due to the biochemical destruction of proteins. Therefore, these methods are not practical for routine sample analysis since proteins are quite unstable after the animal death, they are thermolabile and additionally, they are variable depending on the tissue type, age and status of specimen.

During the 20th century, as an alternative to protein analysis, DNA based methods for species identification have been developed. The advantages of DNA based methods are in the better resistance and theromostablility of the DNA molecule compared to proteins, and by using the polymerase chain reaction (PCR) small fragments of DNA can be amplified providing sufficient information for species identification (Lenstra 2003). Additionally, DNA can be recovered from almost any substrate, since it is present in all cells of an organism. Due to the development of molecular biology, identification of fish species is possible from most tissues including muscles, fins or blood.

There are a few molecular techniques that have been mostly used during the past decade for fish identification as RFLP (restriction fragment length polymorphism), PCR sequencing (known also as FINS – forensically informative nucleotide sequencing) and PCR specific primers. More recently, two methods have developed, real-time PCR and microarray, both used in gene expression research as well as in species identification (Teletchea 2009).

In fish breeding programs, DNA based methods are used for family tracking, pedigree determination of individuals, identification of lineages, identification of markers and loci that are responsible for economically important traits as growth, survivorship, sexual maturity, disease resistance, deformities and nutrient utilization (Davis & Hetzel 2000).

1.1 Application of Restriction fragment length polymorphism (RFLP) in aquaculture and fishery

RFLP is a quite simple, robust and relatively cheaper method (Aranishi 2005). It is based on the polymorphisms in the lengths of particular restriction fragments of the geneticcode (Rasmunsen et al. 2008). The target genes are amplified with PCR and then cut by specific endonucleases on a few smaller fragments, different in size, that are used for identification of fish lineages or species (Liu & Cordes 2004). Different fragments are separated by agarose gel electrophoresis. The gel is then exposed to UV light and with the gel documentation system (transluminator) is photographed. The lengths of the fragments are measured by using the DNA standard (ladder).

This method is particularly useful for the identification of genetic distance between different populations of a species coming from a wider geographic region. It is of key importance that the fish populations under research are highly polymorphic, meaning that they have not been frequently crossbred.

For identification of low polymorphic fish population, other methods mentioned above, or microsatellites are used.

1.2 Application of real-time PCR (qPCR) in aquaculture

A significant part of research in fishery and aquaculture focuses on gene expression as an excellent method for understanding physiological processes in aquatic animals, mainly fish (Overturf 2009). Evaluation of genes that play significant roles in different biological processes as embryonic and adult growth, efficiency in nutrient utilization, disease resistance and other are used in many research programs in fishery and aquaculture (Nilsen & Pavey 2010). Provided by numerous investigations, the growing knowledge in this area has enabled better understanding of fish physiology in aquaculture aiming to achieve economical feasibility and sustainability at the same time.

Quantitative real-time PCR (qPCR) is one of the most accurate methods for gene expression analysis. It is very precise, sensitive, shows good reproducibility, as well as relatively fast results (Derveaux et al. 2010). Although this method is quite easy to use, it is very important to maintain a high level of quality control throughout the entire process. Imprecision in the protocol application can significantly influence the accuracy of results and affect the credibility of research conclusions.

Traditional end-point PCR provides only information on the presence or absence of a specific genetic product, while the advantage of quantitative real-time PCR (qPCR) is in the measurement of the number of copies and the detection of small differences between samples during the reaction.

In real-time qPCR, the detection of PCR products is provided by the presence of a fluorescent dye that binds to the double stranded DNA. The registered amount of fluorescence signal in the qPCR machine responds to the amount of amplified product during every amplification cycle. The most commonly used fluorescent dye that binds to the DNA molecule is SYBR Green I. This non-specific dye binds to the double stranded DNA molecule whose fluorescence intensity in increased up to 1000 times during the binding. Ct or threshold cycle is the value showing the number of amplification cycles during which the fluorescent signal reaches the threshold level (level above the background fluorescence). Lower number of cycles i. e. Ct values lower <29 shows higher amount of the product.

The relative gene expression method is based on the differences in the level of expression in target genes (gene of interest) compared to one or several reference genes (genes whose expression is not affected by the applied treatment) (Pfaffl 2006). To identify the reference gene it first has to be checked how is it expressed. There are several on line programs to evaluate which gene is best to use as reference, e.g. http://fulxie.0fees.us/. The Ct values of the gene are added and the program calculates the most stable gene.

For calculating the gene expression there are different mathematical models. One that is frequently used is the delta delta Ct method:

2^ (-ddCt), where ddCt = (Ct target gen treated– Ctreference gen treated) – (Ct control target gen– Ct control reference gen)

Aquacarp laboratory equipment:

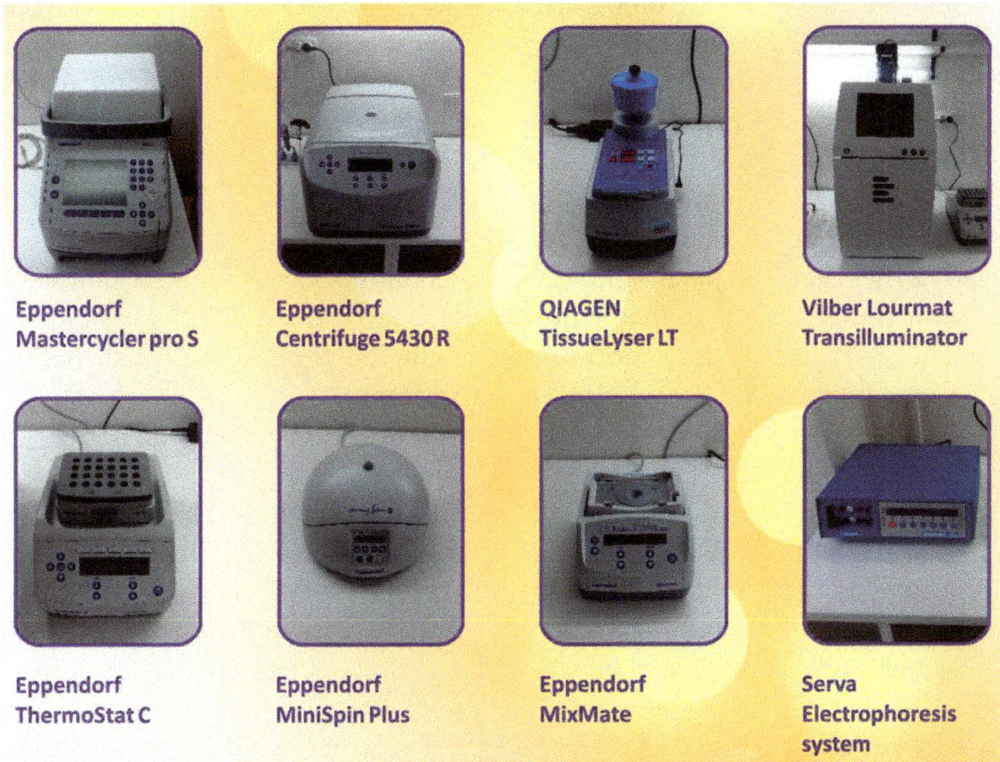

2 Materials, Methods and Notes

*2.1 Determination of genetic lineages of highly polymorphic fish species, Brown trout (*Salmo trutta*), by PCR-RFLP-based analysis*

In this method two genes were tested, one nuclear (gene for lactate dehydrogenase) and one mitochondrial gene (gene for the control region of the mitochondrial DNA).

2.1.1 DNA isolation

For DNA extraction, a commercial kit (ZR Genomic DNA™-Tissue MiniPrep Kit, Zymo Research, Irvine, USA) was used and following procedure is listed according to manufacturer`s instruction:

- 25 mg of fish fin is placed in eppendorf tube of 1.5 ml
- Add to every tube:
  H2O 95 µl
  2X Digestion Buffer 95 µl
  Proteinase K 10 µl

- Vortex the mix and incubate in the thermostat at 55°C 1 do 3 hours (optionally it can stay during thenight). One hour after the incubation shake the tube a couple of times to allow proper digestionof the fin.
- After the incubation period, add 700 µl Genomic Lysis Buffer and vortex the mix thoroughly. Centrifugeat 10,000 × g 1 min to eliminate the cell debris (1 g = 1 RCF).
- Put the Spin Column (Zymo-Spin™ IIC Column) into a 2 ml eppendorf tube without a cap (collection tube)
- Transfer the supernatant into the Spin Column. Centrifuge at 10,000 × g 1 min (DNA binding to themembrane).
- Transfer to a new collection tube and add 200 µl DNA Pre-Wash Buffer. Centrifuge at 10,000 x g 1 min(DNA I washing).
- Add 400 µl g-DNA Wash Buffer into the column. Centrifuge at 10,000 × g 1 min (DNA II washing).
- Transfer to a new column, a regular clean 1.5 ml eppendorf tube with a cap. Add ≥50 µl DNA ElutionBuffer or Milli-Q water (e.g. if 25 mg of fin tissue is used add 200 µl of liquid) in the spin column.
- Incubate for 2–5 min. at room temperature than centrifuge at maximal speed for 30 seconds to eluate the DNA.

Eluated DNA can be used immediately or can be kept at ≤-20°C and used later. If the isolated DNA will be used in the next few weeks, it can be kept in the refrigerator.

2.1.2 Primer selection

For CRmtDNA gene 28RIBa and CytR primers were used and for LDH gene Ldhxon3F and Ldhxon4F (Table 1). These primers were selected since they proved to be very useful in distinguishing trout of Atlantic and Danubian lineages in previous studies (McMeel et al. 2001, Maric et al. 2010).

Primers are received in lyophilized form. They should be dissolved with MilliQ water adding 10x µl more water into the total number of primer nmol: e.g. primer 28RIBa has in total 31.6 nmol, it should be added 31.6 x 10 µl = 316 µl water

Primer name	Sequence	Gene
Ldhxon3F	GGCAGCCTCTTCCTCAAAACGCCCAA	LDH
Ldhxon4R	CAACCTGCTCTCTCCCTCCTGCTGACGAA	LDH
28Riba	CACCCTTAACTCCCAAAGCTAAG	CR mtDNA
CytR	GTGTTATGCTTTAGTTAAGC	CR mtDNA

Table 1: Primers and their sequences used in the study (T.-K. Knutsdatter Østbye).

The working primer solution is made by dissolving the solution 10x (10 μl of solution and 90 μl of water is added into the ependorf tube). Both solutions are kept in the freezer.

2.1.3 PCR-RFLP protocol for control region of mRNA and nuclear gene coding for lactate dehydrogenase

For the PCR reaction, first the PCR Mix should be made. All the reagents, except Taq polymerase, should be taken out of the freezer and warmed up at room temperature. While waiting for the warm up of reagents, put the eppendorf tubes (0.2 ml) in the cooling stand and add 1 μl of eluated DNK into every tube. Number of PCR tubes depend on the number of fish samples (10 fish – 10 PCR eppendorf tubes). The PCR mix is made in one eppendorf tube of 1.5 ml kept in the cooling stand.

The amount for 10 reactions is (if less or more reaction done, this amount should be recalculated):

171.5 μl MilliQ H_2O
25 μl buffer A
12.5 μl $MgCl_2$
10 μl dNTP
10 μl of dissolved forward primer
10 μl of dissolved reverse primer
1.5 μl Taq polymerase

Polymerase is added at the end, very quickly, taking care that it doesn't stay too long at room temperature. Vortex the Mix and add into every PCR eppendorf tube 24 μl:

1 μl DNK + 24 μl Mix-a

The Mix should always be made for one more reaction than the number of samples, to have enough in case something goes wrong e. g. if we analyze DNA of 10 fish, the mix should be made for 11 fish. The PCR eppendorf tubes are kept in the cooling stand. If it starts changing colour, a new cooling stand from the freezer should be taken and tubes transferred to it. Switch on the cycler and wait until the cover reaches 100°C, and then quickly transfer the tubes from the cooling stand into the cycler.

The conditions of the cycler are adopted depending on the gene analyzed.

For the entire controlregion of mitochondrial DNA (CR mtDNA) (1088 bp) the conditions are:

1. step: 94°C – 3 min
2. step: 94°C – 45 sec*
3. step : 54°C – 45 sec*
4. step: 72°C – 1 min 20 sec*

5. step: 72°C – 10 min
6. step: 10°C – ∞

*** steps 2–4 are repeated 32 times

For the nuclear gene coding for lactate dehydrogenase (LDH) (428 bp), the conditions are:

1. step: 94°C – 3 min
2. step: 94°C – 45 sec*
3. step: 62°C – 45 sec*
4. step: 72°C – 1 min*
5. step: 72°C – 10 min
6. step: 10°C – ∞

*** steps 2–4 are repeated 32 times

After the program is finished the PCR product is kept in the refrigerator for a few weeks. If the product should be kept for longer it should be put it the freezer.

2.1.4 Separation of the LDH and CR mtDNA restriction fragment

Before the gel electrophoresis, the agarose gel should be made, cast into the gel tray, and the buffer added. The buffer is an electrolyte that allows the electric current flow.

Preparation of 1.5% agarose gel

The total amount of gel that should be made is 140–150 ml.
For 140 ml:

- measure 210 g of agarose on the analytical balance in a small beaker.
- Transfer the agarose into the Erlenmeyer flask.
- Add 140 ml TBE x0.5 buffer.
- Put the flask into the microwave and cover it with a Petri dish.
- Switch on the maximal hear and let it boil
- When it starts boiling switch off the microwave and shake the agarose so it can dissolve completely. Than let it boil for 10 more seconds.
- When the solution cools down a bit, so it stays liquid, cast it into the tray with the combs (well former templates) and dams (rubber end caps) already in position.
- Wait until the gel cools down and carefully take out the combs.

Preparation of TBE x5 buffer

Tris 54.48 g
Boric acid 27g
EDTA 10 ml 0.5 M solution
Distillate water up to 1000 ml

Put Tris into a small volume of distillate water and mix well. Add Boric acid and mix. These two substances easily mix. Make the the EDTA buffer I and keep it in the refrigerator in a volumetric flask. TBE x5 buffer is made by 10x dilution (mix 100 ml TBE x5 buffer and add 900 ml of distillate water). TBE x0.5 buffer can be kept at room temperature until it is used up.

Note: *Take care not to put the pH electrode into Tris, because it will destroy it.*

Preparation of sample buffer

Buffer blue is used as a carrier for loading the sample on the gel:
Distilled water 1 ml
Glycerol 0.5 ml
A very small amount of Bromophenol Blue (enough to stain the solution)

The mixture separates into two phases, glycerol and water, but after vortexing the mix for 20 min the phases will integrate. The colour of the solution can vary from red (if the pH is low) to blue (if the solution is neutral or higher than 7). The solution should be vortexed every time before using it. Keep it in dark at room temperature.

After the gel cools down transfer it with the tray into the electrophoresis apparatus chamber, take off the rubber end caps and add TBE x0.5 buffer to cover the gel. Cut a piece of parafilm and place 5–20 μl of sample on it (eluated DNA, PCR product or product of restriction-more details at Fig. 1) in droplets, separated by 2–3 cm from one another. Next to the sample droplets, 5 μl of DNA standard (ladder) should be placed. Into every sample droplet 0.5 μl Midori green is added, except into the ladder where 1 μl of Midori green is added. After this, into every droplet 2–3 μl of Buffer Blue is added. Then the parafilm with sampes (droplets) is transferred to the gel and added into every well. Mark the wells. The ladder is added in the first or last well. Adapt the electrical current to be constantly 120 V.

2.1.5 Gel documentation

Switch on the apparatus (Transilluminator) for taking a picture of the gel and click on the button „Live". Enter the USB into the port that is on the right side of the apparatus. Stop the electrical current 45–60 minutes after the electrophoresis, raise the cover and pour out the buffer. The tray with the gel is transferred near the apparatus. Gel is then moved on to the transluminator surface by opening the apparatus door and sliding out the surface so that the transfer of the gel is easier. Gel should be laid smoothly with no air bubbles present under the gel. Push the transluminator surface back into the apparatus, close the door and switch on the UV light (button for UV light). Sometime the gel slides backward so it is advisable to put a piece of plastic underneath. Adjust the contrast and intensity pressing the buttons „+" or „-" until the sufficient quality is reached. Adjust the zoom and focus on the objective manually (objective is positioned on the top of the apparatus). Take a picture of the gel by pressing the button „freeze". If the USB in in the apparatus, the picture will automatically be saved in the folder „Images" on the USB (Fig. 1.). If the folder doesn't exist, the apparatus will automatically create it, and if the USB is not inserted, the picture will be saved in the internal memory of the apparatus. The gel should be photographed immediately, since after few hours, DNA will become invisible.

Digestion results are visible on Fig.1. On the left side of the DNA ladder the restriction digestof 4 fish sampes (A4, A5, P4, P5) were done with the SatI

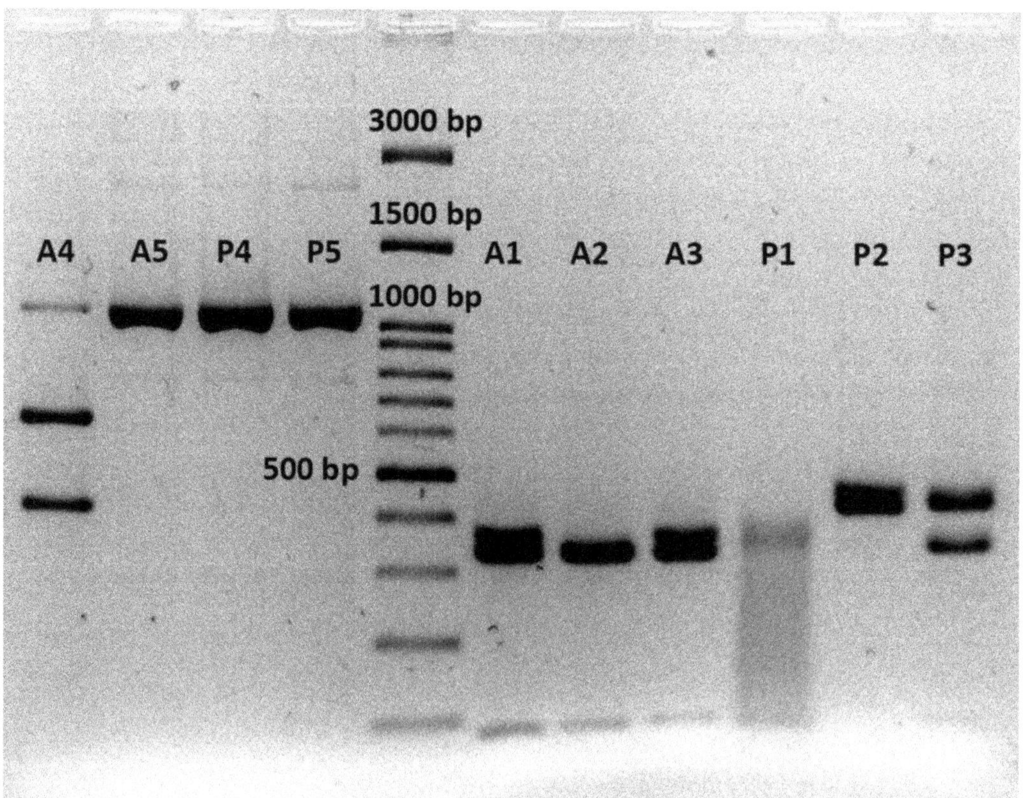

Figure 1: The gel with obtained restriction fragments of brown trout genes after RFLP assay.

enzyme of the control region on the mtDNK (1088 bp). Control region of Danubian lineage remains uncutwhile Atlantic (At) lineage is cut into two fragments (654 and 434bp). On the right of DNA ladder, the restriction digestof 6 fish samples (A1-A3 and P1-P3) is done with the BseLI endonucleasesfor the gene for lactate dehydrogenase (428 bp). Danubian lineage remains uncut whileLDH gene of Atlantic lineage is cut into two fragments (353and 75 bp; fragments of 75 bp were too small to be visible on gels).

*2.2 Expression of the genes in intestines of Atlantic salmon (*Salmo salar*) after various feeding treatments, detected by reverse transcription (RT) qPCR method*

The expression of selected genes was investigated in the intestines of salmon (*Salmo salar*) fed different levels of n-3 fatty acids i. e. eicosapentaenoic acid (EPA) and docosahexaenoic acid (DHA) throughout life. These selected genes are involved in EPA and DHA synthesis, fatty acid oxidation, and synthesis of prostaglandins and immunity.

2.2.1 Sample preparation

Fish were reared in cages and fed pre-diets (5 different types of diets) in the period from 4 to 400g. During the later period fish from 400g to 4kg were fed main diets (3 types). Fish intestines were sampled from the 4kg group and stored at –80°C prior toanalysis. Intestines from six specimens per treatment were selected for the analysis. In total 84 samples.

2.2.2 Total RNA extraction

For Total RNA extraction, a commercial kit (PureLink™ 96 Pro 96 Total RNA Purification Kit, Invitrogen, USA) was used and following procedure is listed according to manufacturer`s instruction:

- Add 1 mL TRIzol™ Reagent (Invitrogen) to 2 mL tubes for Precyllis24 with 2–3 beads.
- Add approximately 20 mg tissue.
- Homogenize the sample in Precyllis24 in centrifuge at 5500 rpm 2x 20 seconds.

At this point samples are processed in batches, in case there is a lot of samples e.g. 24 samples (depending on the type of refrigerated centrifuge) can be further processed while rest have to be stored at –80°C, and processed within a month.

2.2.3 Trizol protocol:

- Incubate the homogenized sample for 5 minutes at room temperature to permit the completedissociation of the nucleoprotein complexes.
- Add 0.2 ml of chloroform per 1 ml of Trizol Reagent. Cap sample tubes securely.
- Shake tubes vigorously by hand for 15 seconds.
- Incubate the samples at room temperature for 2 to 3 minutes.
- Centrifuge the samples at no more than 12,000 × g for 15 minutes in the refrigerated centrifuge at 4°C (Fig.2 right).

Following centrifugation, the mixture separates into a lower red, phenol-chloroform phase, an interphase, and a colourless upper aqueous phase. RNA remains exclusively in the aqueous phase. The volume of the aqueous phase is about 60% of the volume of Trizol Reagentused for homogenization (Fig.2 left).

2.2.4 Transfer of samples to PureLink™ 96 Well Total RNA Filter Plates

- Transfer 350 uL the colourless upper aqueous phase to new tubes. Add 1× volume of PureLink Pro96 Lysis Buffer, 350 uL.
- Add a volume of 100% ethanol equal to the aqueous phase;x. 350 uL.
- Mix well.
- Place the PureLink™ 96 Well Total RNA Filter Plate on top of a PureLink™ 96 Receiver Plate. Transfer samples to the PureLink™ Well Total RNA Filter Plate (Fig.3).
- Seal the plates with a cover.
- Centrifuge the stacked plates at ≥2,100 × g for 1–2 minutes at room temperature.

Note: *If some of the samples did not go through the columns, remove the cover and spin 1 minute at 2,100 × g.*

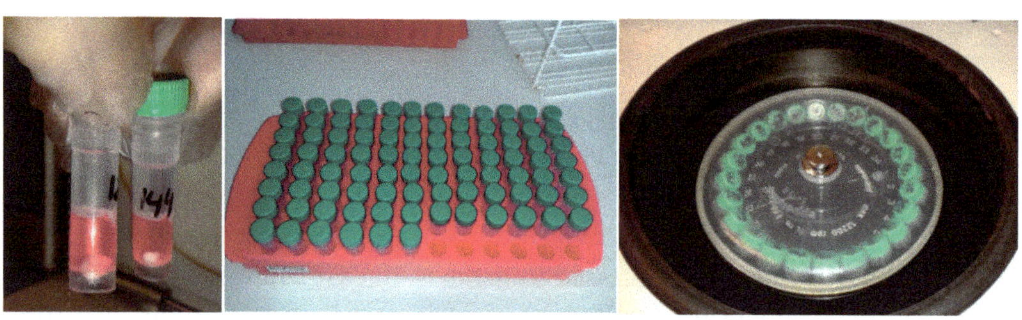

Figure 2: Left-tubes with separated phases, middle – preparation of tubes for tissue homogenization, right – 24 tubes in refrigerated centrifuge.

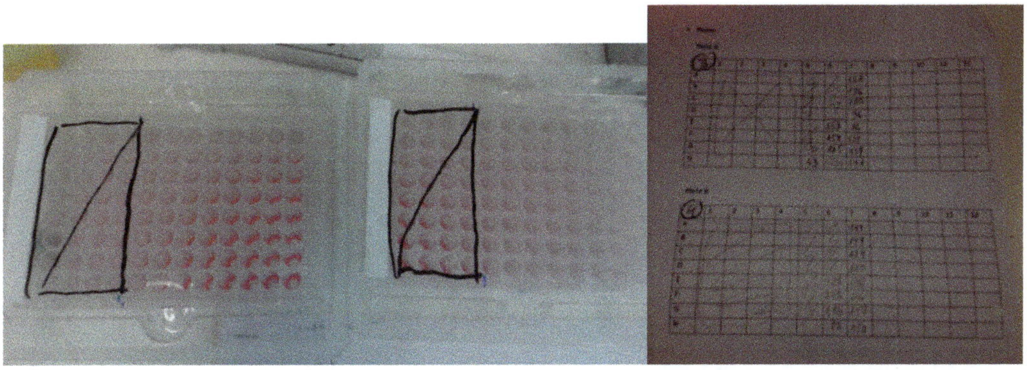

Figure 3: Left – marked PureLink™ 96 Well Total RNA Filter Plates, right – samples numbers marked in the tables indicating which will be used in the PureLink™ 96 Well Total RNA Filter Plates.

	100×:	1×:	16×:	26×:
10 × DNase I buffer	0,8 ml	8 µl	128 µl	208 µl
DNase I	2700 U	9,9 µl	158 µl	257.4 µl
RNase-free water	to 8 ml	62 µl	992 µl	1612 µl

Table 2: Amount of DNase reagents for digestion, right highlighted column shows the amount for 26samples.

- Discard flow-through from the PureLink™ Receiver Plate and place the PureLink™ RNA Filter Plateon top of the PureLink™ Receiver Plate.

2.2.5 On-Column DNase Digestion:

For On-Column DNase Digestion a commercial kit (PureLink™ 96 DNase for use with PureLink™ Kits,On-Column Protocol Only, Invitrogen, USA) was used and following procedure is listed according to manufacturer`s instruction:

Prepare DNase solution depending on the number of samples, e.g. if you have 24 samples always add reagents for two more reactions (Table 2).

***Note:** Make the DNAse solution in a bigger eppendnof tube and keep it on ice until used.*

- Add 500 uL wash buffer I to each sample. Centrifuge at 2100 xg for 2 minutes.
- Add 80uL DNase solution to each sample.
- Incubate for 15 minutes at room temperature

2.2.6 Buffer washing

- Add 500 uL wash buffer I and incubate for 5 minutes
- Centrifuge at 2100 xg for 2 minutes, discard flow-through.
- Add 750 uL wash buffer II – Centrifuge at 2100 xg for 2 minutes, discard flow-through.
- Add 750 uL wash buffer II
- Centrifuge at 2100 xg for 2 minutes, discard flow-through.
- Dry the plate by centrifugation at 2100 xg for 10 minutes.

Note: *Buffer II should be prepared freshly before adding to the samples. Buffer II needs to be diluted 1:5 with 95–100% ethanol; 24×750 uLx 2 (washes)=36 ml (8ml of 5× buffer II+32ml of 96% ethanol)*

2.2.7 RNA Eluation

- Place the filter plate on top of a new PureLink™ Pro 96 Eluation Plate (Fig. 4 left).
- Add approximately 60 uL RNase free water (regularly add 45uL, but more can be added, depending on the type of tissue).
- Incubate for 1 minute at room temperature.
- Centrifuge at 2100 xg for 2 minutes to eluate RNA.
- Transfer eluated RNA to tubes. Work on ice.
- Measure concentration and quality of RNA of every sample with the Nanodrop (Fig.4 right).
- Store the RNA samples at –80°C.

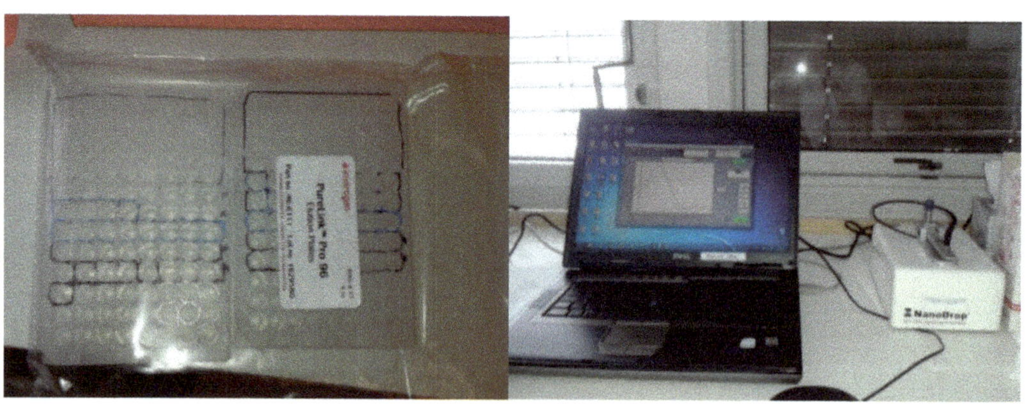

Figure 4: Left PureLink™ Pro Eluation Plates with marked wells that will be used, right Nanodrop spectrophotometer connected to the computer program.

2.2.8 TaqMan qPCR Assay: cDNA synthesis and specific gene amplification

For cDNA synthesis and expression of selected genes TaqMan Reverse Transcription Reagents (Applied Biosystems) with SYBR™ Green (LightCycler® 480SYBRGreen Master I) dye method for labelling and detecting qPCR products was used.

cDNA was made from 1000 ng of RNA. According to the concentration of RNA received from the Nanodrop, calculation is made for every sample by dividing 1000 with the RNA concentration e.g. 1000/536.2= 1.80 µL RNA. In order to calculate the amount of RNA free water that that should be added to every sample, the µL of RNA is subtracted from 7.7 (e.g. 7.7–1.8 = 5.8 µL of RNA free water) for a 20 µL TaqMan reaction (Table 3).

For calculating the amount of Taqman regents for the total number of samples, the number of samples (84) is multiplied with the amount of every TaqMan® reagent (Table 3, right). Than the volume of TaqMan reagent per one

Taqman cDNA synthesis			**Number of samples**
TaqMan® Reverse Transcription Reagents	stock conc.	20 uL reaction (uL)	84
10x TaqMan® RT Buffer	10×	2	168
25mM $MgCl_2$	25 mM	4.4	369.6
10mM dNTP mixture	10 mM	4	336
Oligo $d(T)_{16,\ 50\ uM}$	50 uM	1	84
RNase Inhibitor	20 U/ul	0.4	33.6
Reverse Transcriptase (50 U/µl)	50 U/ul	0.5	42
RNA 250–1000 ng		x	Σ 1033.2
Rnase free water		7.7	
Total		20	

Table 3: TaqMan® Reverse Transcription Reagents and calculation of the amount for the 20 uL reaction, right column amount of TaqMan reagents for 84 samples.

Note: *The first 4 reagens from the table should be kept on ice before using them, while the Rnase inhibitor and Reverse Transcriptase are kept frozen before using.*

sample is calculated by dividing the total volume of reagents for all samples with the number of samples (e. g. 1033.2/84=12.3 μL).

Samples are transferred to a new type of plated and RNA free water and 12.3 μL per sample of Taqman mixture is added. In one well a mixture of RNA samples without enzymes is added to check for the presence of genomic DNA (NTC – non template control) (Table 4. up).

After adding all the parts, the plate are shortly centrifuged (1 min on 2000g), and placed in the PCR machine. The reaction conditions should be 25°C for 10 min, 48°C for 60 min and 95°C for 5 min

After the amplification, 180μL of RNA free water is added to every sample. Then from every sample 10 μL is taken and transferred into one eppendorfer tube to make the "cDNA mix". The "cDNA mix" is used for making a standard curve later.

Next step is transferring the samples from the main plate to the master plates (1 and 2) (Table 4. up). Half and half of samples (42) are transferred to the master plates 1 and 2. During this every sample was duplicated (Table 4, qPCR master plate 1 and master 2).

Using the microdispenser machine, from every master plate, 15 white LightCycler per master plate were copied (in total 30 plates). After this process, every sample contains 4μL of cDNA.

Note: *Always make more copies than number of primers analyzed. (e.g. if 13 primers are analysed, 10 target genes and 3 referent genes, a few more copies are made – 15 plates per master plate, in total 30 new plates (Fig.5).*

Every plate is covered with a special sealing cover, centrifuged for 1min at 2000 rpm, and stored at -20 until used.

Next step is adding chosen primers and SybrGreen fluorescent dye to every sample before placing them into the qPCR machine.

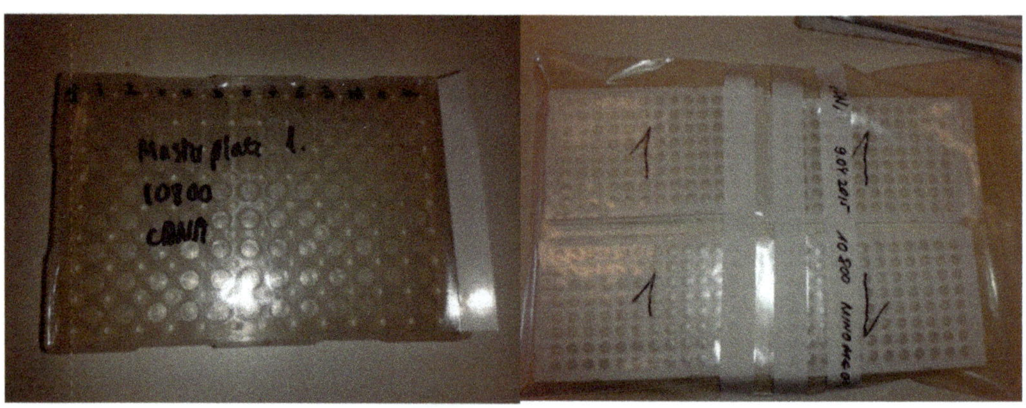

Figure 5: Left master plate 1, right 15 copies of master plate.

cDNA synthesis

	1	2	3	4	5	6	7	8	9	10	11	12
A	1	3	7	8	9	10	11	13	14	16	20	23
B	25	26	27	28	29	30	33	40	43	45	62	72
C	153	86	144	136	128	149	141	108	147	116	82	130
D	128	154	134	140	123	115	139	96	150	110	53	146
E	182	80	163	124	212	88	183	91	159	173	180	221
F	175	170	107	185	122	145	111	168	118	174	169	165
G	54	117	77	126	102	160	161	90	NTC			
H												

qPCR masterplate 1

	1	2	3	4	5	6	7	8	9	10	11	12
A	1	1	3	3	7	7	8	8	9	9	10	10
B	25	25	26	26	27	27	28	28	29	29	30	30
C	153	153	86	86	144	144	136	136	128	128	149	149
D	128	128	154	154	134	134	140	140	123	123	115	115
E	182	182	80	80	163	163	124	124	212	212	88	88
F	175	175	170	170	107	107	185	185	122	122	145	145
G	54	54	117	117	77	77	126	126	102	102	160	160
H	cDNA mix	cDNA mix	cDNA mix									

qPCR masterplate 2

	1	2	3	4	5	6	7	8	9	10	11	12
A	11	11	13	13	14	14	16	16	20	20	23	23
B	33	33	40	40	43	43	45	45	62	62	72	72
C	141	141	108	108	147	147	116	116	82	82	130	130
D	139	139	96	96	150	150	110	110	53	53	146	146
E	183	183	91	91	159	159	173	173	180	180	221	221
F	111	111	168	168	118	118	174	174	169	169	165	165
G	161	161	90	90	NTC	NTC						
H	cDNA mix	cDNA mix	cDNA mix									

Table 4: Upper table, main plate used for making the two master plates (qPCR masterplate 1 i 2), middle and lower table.

For the qPCR reaction mix use:
0.5 μL reverse primer
0.5 μL forward primer
5 μL Sybrgreen dye
4 μL cDNA (amount in every sample)

Diluted primers are kept in the freezer. Defrost1–2 minutes at room temperature before use. Initially, primers are delivered in lyophilized form and are dissolved with RNA free water. The amount of primer needed should be calculated according to the total number of reactions. If 96 well plates (96 samples) are used than 0.5 μL × 96=48 × 2 (if there is two master plates)= 96μL of primer. Into this, 10× more RNase free water is added – 960 μL.

Selected target genes were ACO, COX1, COX2, d5d, d6fad_a, d6fad_b, Elovl2, Elovl5b, nrf, nfkb. Referent gene were: etif and Ef1a (Table 5).

Every primer is made in a separate eppendorf tube marked at the beginning (e.g. COX1 r and COX1 f) (Fig.6).

Dissolved primers are kept on ice before transferring to the samples. SybrGreen is kept in the freezer until the package is opened. An open package is saved in the fridge until it's used up. It should be used within one week. The amount of SybrGreen is calculated for the total number of reactions 5 μL × 96=480μL.

Into every sample 6 μL of dissolved primer and Sybrgreen reagent is added preferably using the multistep pipette, especially when having plates with 96 or more wells. The plates are sealed with a special type of translucent protective foil for qPCR (LightCycler sealing foil) and centrifuged 1 min at 2000 rpm. The plates are then placed in the LightCycler (qPCR machine) and runned using the chosen type of program (Fig. 7).

The qPCR reaction was run on a LightCycler™480 (Roche Diagnostics Gmbh, Germany) under the following conditions: Preincubation at 95°C for 5 minutes, amplification with 45 cycles at 95°C for 15 seconds and 60°C for 1 minutes, melting curve at 95°C for 5 seconds and 65°C for 1 minutes, cooling at 40°C for 10 seconds.

Figure 6: Left reverse and forward primer COX1 (marked with register numebrs), right marked eppendorf tube for dissolved primer COX 1r.

Primer name	Sequence
Ssa elong 2F1	CGGGTACAAAATGTGCTGGT
Ssa elong 2R1	TCTGTTTGCCGATAGCCATT
Ssa ACO F1	CCTTCATTGTACCTCTCCGCA
Ssa ACO R1	CATTTCAACCTCATCAAAGCCAA
Ssa d5d F2	GCTTGAGCCCGATGGAGG
Ssa d5d R2	CAAGATGGAATGCGGAAAATG
Ssa d6d A F3	TCCCCAGACGTTTGTGTCAGATGC
Ssa d6d A R3	GCTTTGGATCCCCCATTAGTTCCTG
Ssa d6d B/C R2	CACAAACGTCTAGGAAATGTCC
Ssa d6d B F3	TGACCATGTGGAGAGTGAGGG
Ssa d6d B R3	AACTTTTGTAGTACGTGATTCCAGCT
Ssa d6d C F2	TGAAGAAAGGCATCATTGATGTTG
Ssa EF1a F	CACCACCGGCCATCTGATCTACAA
Ssa EF1a R	TCAGCAGCCTCCTTCTCGAACTTC
Ssa etif3 F1	CAGGATGTTGTTGCTGGATGGG
Ssa Etif3 R1	ACCCAACTGGGCAGGTCAAGA
Ssa nrf F2	CCGGACTCCTCGCCTTCGGA
Ssa nrf R2	GTGGATAGTTGGCTTGTCCCTTCGT
Ssa nfkb F1	CAGCGTCCTACCAGGCTAAAGAGAT
Ssa nfkb R1	GCTGTTCGATCCATCCGCACTAT
Ssa Elovl5b F2	GCAACCTTGACCCAAACAGG
Ssa Elovl5b R2	CCTTGTCTCTACGCAAGGGA
Ssa cox2 F2	CCCCCGACTTACAATGCTGA
Ssa cox2 R2	GCGGTTCCCATAGGTGTAGG

Table 5: The primers and sequences used in the study (T.-K. Knutsdatter Østbye).

Figure 7: Left – plates ready for adding the primers and SybrGreen-a, middle – Roche™ LightCycler™ 480, right – monitoring the process of amplification on the LightCycler computer program.

Note: *Be careful while putting the solution in the wells, since this small volume, can jump out of the well. Best is to bring the pipette tip next to the wall lower in the well, but not touching. If a droplet "jumps" out of the well, take it a away with a ordinary pipette and tip, and add once more a droplet from the multistep pipette. Take a note on the exact sample where we made a mistake.*

After all plates are runned in the qPCR machine, the data for Ct values for all reactions are taken from the LyghtCycler software (can be transfered to excel) and are further processed using chosen mathematical model for relative gene expression.

3 Acknowledgements

This work was funded by EU Commission project AREA, no. 316004. These analyses are a part of an ongoing project at Nofima, National food Institute in Norway, As, at the Biotechnology laboratory. We are grateful to Bente Ruyter, Tone-Kari Knutsdatter Østbye and Marta Bou Mira for providing samples and tremendous help during the analyses.

4 References

Aranishi, F. (2005). Rapid PCR-RFLP method for discrimination of imported and domestic mackerel. *Marine Biotechnology*, 7(6), 571–575. DOI: https://doi.org/10.1016/j.aquaculture.2004.05.027

Davis, G. P., & Hetzel, D. G. (2000). Integrating molecular genetic technology with traditional approaches for genetic improvement in aquaculture species. *Aquaculture Research*, 31, 3–10. DOI: https://doi.org/10.1046/j.1365-2109.2000.00438.x

Derveaux, S., Vandesompele, J., & Hellemans, J. (2010). How to do successful gene expression analysis using real-time PCR. *Methods*, 50, 227–230. DOI: https://doi.org/10.1016/j.ymeth.2009.11.001

Granadeiro, J. P., & Silva, M. A. (2000). The use of otoliths and vertebraein the identification and size-estimation of fish inpredator-prey studies. *Cybium*, 24, 383–393.

Hubalkova. Z., Kralik, P., Tremlova, B., & Rencova, E. (2007). Methods of gadoid fish species identification in food and their economicimpact in the Czech Republic: a review. *Veterinarni Medicina,* 52,273–292.

Kvasnicka, F. (2005). Capillary electrophoresis in food authenticity. *Journal of Separation Science,* 28,813–825, DOI: https://doi.org/10.1002/jssc.200500054

Lenstra, J. A. (2003). DNA methods for identifying plant and animal species in food. *In*: M.Lees (Ed.), *Food authenticity and traceability* (pp. 34–53). Cambridge, U.K.,Woodhead Publishing Ltd.

Liu, ZJ., & Cordes, J. F. (2004). DNA marker technologies and their applications in aquaculture genetics. *Aquaculture*, 238(1–4), 1–37.

Marić, S., Simonović, P., & Razpet, A. (2010). Genetic characterization of broodstock brown troutfrom Bled fish-farm, Slovenia. *Periodicum Biologorum*, 112,145–148. ISSN 0031-5362

McMeel, M. O., HOEY, M. E., & Ferguson, A. (2001). Partial nucleotide sequences, and routine typing by polymerase chain reaction–restriction fragment length polymorphism, of the brown trout (*Salmo trutta*) lactate dehydrogenase, LDH-C1*90 and *100 alleles. *Molecular Ecology*, 10, 29–34. DOI: https://doi.org/10.1046/j.1365-294X.2001.01166.x

Nielsen, J. L. & Pavey, S. A. (2010). Perspectives: Gene expression in fisheries management. *Current Zoology*. 56(1), 157–174.

Overturf, K. (2009) Quantitive PCR. In K. Overturf (Ed.), *Molecular Research in Aquaculture* (pp 39–56), Wiley-Blackwell, Oxford, UK. DOI: https://doi.org/10.1002/9780813807379

Pfaffl, M. W. (2006). Relative quantification. In T. Dorak (Ed.) *Real-time PCR* (pp. 63–82). Taylor & Francis Group.

Rasmussen, R. S., & Morrissey, M. T. (2008). DNA-Based Methods for the Identification of Commercial Fish and Seafood Species. *Comprehensive reviews in food science and food safety*, 7, 280–295. DOI: https://doi.org/10.1111/j.1541-4337.2008.00046.x

Strauss, R. E. & Bond, C. E. (1990). Taxonomic methods: morphology. In C.B. Schreck, P.B. Moyle (Eds.) Methods for fish biology (pp. 109–140). American Fisheries Society, Maryland.

Teletchea, F. (2009). Molecular identification methods of fish species: reassessment and possible applications. *Reviews in Fish Biology and Fisheries*, 19, 265–293. DOI: https://doi.org/10.1007/s11160-009-9107-4

Useful links to the topics of this chapter:

http://www.gene-quantification.com/real-time-pcr-handbook-life-technologies-update-flr.pdf

http://www.bio-rad.com/webroot/web/pdf/lsr/literature/Bulletin_5279.pdf

http://www.gene-quantification.info/

http://fulxie.0fees.us/

http://eur-lex.europa.eu/legal-content/EN/TXT/?uri=URISERV%3Al66002

PART II

Raman microscopy/ spectroscopy in agricultural sciences and food technology

Introduction to Raman Microscopy/ Spectroscopy

Dejan Lazić

Abstract

Raman microscopy/spectroscopy is a non-destructive technique that affords detection of vibrational, rotational or other low frequency excitations of the observed system. Vibration will be active in the Raman spectrum when it causes a change in polarizability of molecule. Characteristics of Raman spectra depend on the mass of the atoms, their geometrical arrangement in the molecule, as well as the intensity of the forces that connect these atoms. Coupling Raman spectroscopy with microscopy afforded obtaining precise morphological and structural images of the observed sample. A large number of papers has been published in the field of physico-chemical characterization of materials using Raman microspectroscopy. Its use has slowly propagated into biological sciences as a method for quality control of pharmaceutical products. Application of Raman microspectroscopy will expand as researchers of biological systems become more aware of the possibilities of this method.

Spectroscopy is the field that deals with the study of radiation after its interaction with matter. Such radiation may have electromagnetic or mechanical properties or it might be composed of particles. In order for frequency spectrum

How to cite this book chapter:
Lazić, D. 2019. Introduction to Raman Microscopy/Spectroscopy. In: Vucelić Radović, B., Lazić, D. and Nikšić, M. (eds.) *Application of Molecular Methods and Raman Microscopy/Spectroscopy in Agricultural Sciences and Food Technology*, Pp. 143–150. London: Ubiquity Press. DOI: https://doi.org/10.5334/bbj.i. License: CC-BY 4.0

of electromagnetic radiation to be used, interval of the frequencies can be between frequencies of radio waves to the frequency of γ radiation, whereas for the radiation made up from particles interaction with electrons, neutrons, muons, and in some cases the entire atoms or ions can be used. Information on the matter is obtained from the spectrum that occurs after interaction of radiation with the matter (Kuzmany 2009).

Due to the possibility of determining physical and chemical properties of a substance, Raman microscopy is used in chemistry, medicine, astronomy, metallurgy.

The methods used in spectroscopy vary depending on which substance is used, an atom, molecule or substance, and of the properties of matter that are being tested (http://web.mit.edu/spectroscopy/history/history-classical.html). Thus, spectroscopy as a discipline can be divided into several sub-disciplines, according to:

- The type of radiation – it'sfrequency (radio, microwave, and ahercna, infrared, visible, ultraviolet, X-ray, gamma)
- Level of interaction (nuclear, atomic, molecular, aggregation)
- Inventor (Raman, Mesbauerova)
- Technique (Fourier, resonance, laser, coherent, emission, absorbent)

Based on the phenomenon that causes coupling of electromagnetic radiation spectroscopy can be divided into:

- rotational spectroscopy
- electron spectroscopy
- vibrational spectroscopy

(Http://web.mit.edu/spectroscopy/history/history-classical.html)

Vibrational spectroscopy is the branch of spectroscopy that studies the spectra, which are caused by vibrations of molecules. This includes infrared spectroscopy and Raman spectroscopy. It is used for identification of molecules based on their atomic vibrations, and periodic changes in the angles and distances of atoms.

Vibrational spectroscopic techniques can be based on the absorption or dispersion of electromagnetic radiation, and therefore we can distinguish between infrared and Raman spectroscopy.

All molecules vibrate. In contrast to the vibrations of macroscopic bodies vibrations of molecules are quantum. Molecules can vibrate with very specific amplitudes which correspond to the vibrational energy levels. The spectroscopic transition can occur only if the energy of absorbed or emitted photon is equal to the difference between two vibrational energy levels (Bohr condition) or if the difference between energy of scattered and the energy of the incoming photon equals to the energy difference between levels (Figure 1). If vibrations of molecule cause change in the dipole moment of the molecule, coupling between such vibrations with vibrations of electromagnetic radiation will happen which causes either absorption or emission of radiation. Such vibrations will be active in the infrared spectrum (Crouch 2007). Vibration will be active in the Raman spectrum when it causes a change in polarizability of molecule.

The study of the spectrum of electromagnetic radiation is carried out using a spectroscope i.e. spectrometer. The analysis provides data on the presence of different types of atoms and molecules in matter.

Since the molecule represents a system of atoms which are linked together by certain forces, as a result of this resilient connection it is possible for the atoms within the molecule to oscillate. Oscillation of atoms and the changes occurring as a result of these oscillations are described by vibrational spectra. These characteristics depend on the mass of the atoms, their geometrical arrangement in the molecule, as well as the intensity of the forces that connect these atoms. Substances containing a large number of molecules such as crystals act like super molecules and in them individual oscillations of molecules can couple and intereact with each other. The exact explanation of these phenomena is not possible; therefore, certain approximations must be introduced. One of the introduced approximations is neglecting mutual interactions of molecules. As any substance or molecule has its own unique characteristics their vibrational spectra will vary. Using Raman spectroscopy one can identify the type of unknown molecule. Also, it is possible to determine molecular structure and to calculate other characteristics of the molecules from their vibrational spectra. That is why today this kind of spectroscopy plays an important role in science.

1 Raman spectroscopy

Sir Chandrasekhara Venkata Raman (1888 -1970) was an Indian physicist whose work largely influenced the development of physics. In 1930 he won the Nobel Prize for Physics for explaining the effect of light changing its wavelengths as it passed through the transparent material (Raman scattering) (http://en.wikipedia.org/wiki/C._V._Raman).

Inelastic reflections of light in interaction with matter theoretically predicted Brillouin in 1922. The first to observe it in experiments were Raman and Krishnan in 1928. For the light source they used focused and filtered beam of sunlight, the sample was liquid in the bucket, a detector was an eye. The same principle applies today. During development of the method as a light source different lamps, lasers were used and today laser diodes prevail, while photomultipliers or CCD cameras are used as the detectors.

Raman spectroscopy uses inelastic scattering of monochromatic light (usually a laser). Raman scattering is used to gather spectroscopic data. Inelastic scattering means that the frequency of monochromatic photons in the light changes after its interaction with the sample. The electromagnetic radiation that is scattered by a molecule contains two components which come from the vibration or the rotation of the molecules (Schrader 1995). Photon that exits the system does not have the same energy as a photon entering the system. Raman scattering system receives or loses power. The difference in the energy of the system before and after scattering corresponds to the difference between the energy of the incoming and outgoing photons. If the outgoing photon has lower energy of incoming photons then the system will recieve energy through scattering. This type of scattering is called Stokes scattering, in the opposite case, the scattering is called anti-Stokes (Figure 1).

Energy of spectroscopic transfer is determined by the energy difference of Stokes or anti-Stokes band and the wavelength of incident monochromatic radiation. As the Stokes and anti-Stokes bands are of much lower intensity than scattered Rayleigh* radiation it is necessary to apply a high-intensity radiation source (Schrader 1995).

*Rayleigh scattering in the atmosphere is due to the scattering of light on air molecules (whose dimensions are much smaller than the wavelength of the incident radiation) that are moving chaotically, thus it is the result of the fluctuations in air density (leading to different values of refractive index in the areas of different densities).

Stokes and anti-Stokes scattering depends on the change in polarizability of molecules in time and in the Raman spectrum only the vibrations and rotations that modify the polarizability of molecules can be observed. Therefore, in the Raman and infrared spectra, the same bands usually have different intensities, and sometimes they are in one of these spectra completely invisible. Therefore, infrared and Raman spectroscopy are considered complementary (Schrader 1995)

Raman spectroscopy has a great advantage over infrared spectroscopy because Raman spectra can be recorded in aqueous solutions. Raman spectroscopy can be also used for the study of solid and gas samples. Around 99.999% of all photons in spontaneous Raman spectroscopy belong to elastic Rayleigh scattering category. This type of scattering is useless for practical application for molecular characterization. Only 0.001% of the incident monochromatic light produces inelastic Raman signal of frequency Vo ± Vm. Spontaneous Raman scattering is very weak and it is necessary to take specific measures in order to differentiate it from the prevailing Rayleigh scattering.

Raman spectroscopy system typically consists of four main components: the source of radiation that is usually a continuous laser type system for illuminating the sample, the system for separating wavelengths and the systems for detection and signal processing (Ferraro 2002).

Raman spectrophotometers use the source of monochromatic radiation that is directed at the sample. Radiation, scattered at an angle (usually 90°) is leaded to the monochromator which allows radiation of only one wavelength to go out. Spectrum is obtained by scanning in the wavelengths around the wavelength of the radiation source.

Raman spectra can be:

- rotational
- rotaional vibrational and
- electron (rarely)

Raman effect is based on inelastic scattering of monochromatic incident radiation (laser radiation) from a ultraviolet, visible or near IR part of spectra.

- When light passes through a transparent material, the largest part of the light is scattered in all directions.
- Majority of the light, consisting of photons whose energy is not converted, is known as Rayleigh scattering.
- The much smaller part of the incident radiation having a change in energy is known as Raman scattering.
- The difference in energy is directly dependent on the molecular species present in the assay system.

If the incident radiation has a discrete spectrum then the spectrum of the scattered radiation in the three components occurs:

- radiation of the same frequency as the incident radiation (Rayleigh scattering)
- radiation of a higher frequency of the incident radiation (Stokes ANTI-RADIATION) and
- radiation of lower frequency of the incident radiation (Stokes RADIATION)

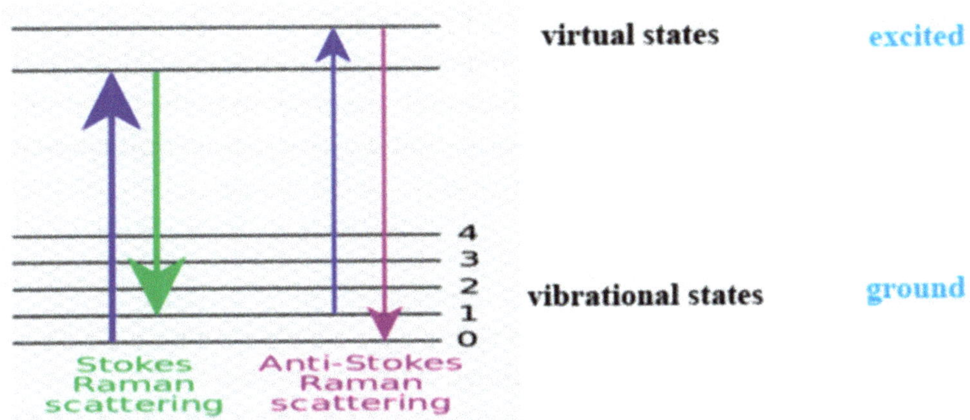

Figure 1: Stokes and anti-Stokess Raman scattering.

1.1 Surface-enhanced Raman spectroscopy (SERS) and surface-enhanced resonance spectroscopy

Spectroscopy of surface enhanced Raman scattering (SERS) is a method that has been developing in the last thirty years, and owing to its very high sensitivity it found application in numerous areas such as chemical and biochemical analysis, medicine, forensics and archeology (McNay et al. 2011, Schlüker et al. 2011). SERS spectroscopy provides vibrational spectrum of the molecule that has been adsorbed onto metal nanostructured surface. Practical use of SERS is difficult due to the formation of spectrum that is very difficult to interpret. It is extremely necessary to take into account physical and chemical factors influencing appearance and intensity of lines in the spectrum. To overcome these complications a type of SERS that uses amplification effect due to adsorption on the surface and Raman resonance was developed. It gives signal intensity up to even 10^{14} (Stiles et al 2008; Schatz and Van Duyne, 2002). The most important reason for the development of SERS technique is its applicability in various fields of natural sciences. By increasing intensity of scattered radiation on a metal substrate molecule, sensitivity of Raman spectroscopy is also increased and allows structural analysis of molecules based on vibrational spectrum. Since spectroscopy of surface enhanced Raman scattering has exceptional sensitivity and provides information on the structure of molecules, it is an ideal technique for examination of various chemical and biological systems. An additional advantage of the SERS spectroscopy is fluorescence quenching near the metal surface due to transfer of energy from the molecule to the metal (Schlüker 2011).

1.2 Raman microspectroscopy

Coupling Raman spectroscopy with microscopy has become significant during 70-ies. In addition to its classical use, use of the electron beam in microscopy gave precise morphological and structural images of the observed sample. Scanning electron microscopy, x-rey microanalysis and analytical electron microscopy give precise details of the elemental composition of the tested substance but these methods don't give information on structural coordination as can be obtained by Raman microspectroscopy. A large number of papers have been published in the field of physico-chemical characterization of materials using Raman microspectroscopy. Its use has slowly propagated into biological sciences as a method for quality control of pharmaceutical products. Application of Raman microspectroscopy will expand as researchers of biological systems become more aware of the possibilities of this method.

Although Raman microspectroscopy is a relatively new method, its popularity due to simple procedure of sample preparation and high number of data available from one spectrum has lead to development of a high number of specific protocols. Majority of protocols is focused on enhancing specifity and sensitivity of the method. This is done by modifying sample preparation, changing the parameters etc. (Butler et al. 2016). In order to facilitate data manipulation and to allow classification of samples based on their „ spectral fingerprint“, new softwares and subprograms in already existing programs were developed (Table 1).

Software	Website	Licence
CytoSpec	http://www.cytospec.com/ftir.php	Comercial
ImageLab	http://www.imagelab.at/en_home.html	Comercial
Biodata Toolbox (in MATLAB)	http://www.mathworks.com/matlabcentral/fileexchange/22068-biodata-toolbox	Open acess
Extended Multiplicative Signal Correction (EMSC) Toolbox	http://www.models.life.ku.dk/emsctoolbox	Open acess
IrootLab	https://code.google.com/p/irootlab/	Open acess
Multivariate Image Analysis (MIA) Toolbox	http://www.eigenvector.com/software/mia_toolbox.htm	Open acess
Multivariate Curve Resolution– Alternating Least Squares (MCR-ALS) Toolbox	http://www.cid.csic.es/homes/rtaqam/tmp/WEB_MCR/welcome.htm	Comercial
PLS Toolbox (in MATLAB)	http://www.eigenvector.com/software/pls_toolbox.htm	Open acess
Raman Processing Program (in MATLAB-a)	http://cares.wayne.edu/rp/	Comercial
Origin for Spectroscopy	http://www.originlab.com/index.aspx?go=Solutions/Applications/Spectroscopy	Comercial
PeakFit	https://systatsoftware.com/products/peakfit/	Comercial
PyChem (in Python programe)	http://pychem.sourceforge.net/	Open acess
PyVib2(in Python programe)	http://pyvib2.sourceforge.net/	Open acess
HyperSpec (in R programe)	http://hyperspec.r-forge.r-project.org/	Open acess
The Unscrambler X (in R programe)	http://www.camo.com/	Comercial

Table 1: Softwares for data manipulation and classification of Raman spectra (adopted from Butler et al. 2016).

2 Acknowledgments

Thanks to the effort of a group of professors of Faculty of Agriculture in Belgrade and with the support of the EU, with funds received through AREA project Raman microscope was purchased and is used for research in agriculture. Researchers from our Faculty had the opportunity to familiarize themselves with the method through trainings at the Institute of Physical Chemistry of the University of Jena (Germany), where they gathered top researchers in the field of application of Raman spectroscopy in biological systems. Here are presented their experiences, impressions, messages to future researchers in this field.

3 References

Butler H.J., Ashton L., Bird B., Cinque G., Curtis K., Dorney J., Esmonde-White K., Fullwood N.J., Gardner B., Martin-Hirsch P.L., Walsh M.J., McAinsh M.R., Stone N. & Martin F.L. (2016). Using Raman spectroscopy to characterize biological materials. *Nature Protocols,* 11(4), 664–687. DOI: https://doi.org/10.1038/nprot.2016.036

Kuzmany, H. (2009). Solid-State Spectroscopy. 2nd edn. Springer-Verlag, Berlin. DOI: https://doi.org/10.1007/978-3-642-01479-6

MIT Spectroscopy: The Era of Classical Spectroscopy. Retrieved from http://web.mit.edu/spectroscopy/history/history-classical.html

Crouch, A. (2007) Principles of instrumental analysis. Thomson Brooks/Cole, Australlia. DOI: https://doi.org/10.1021/ed069pA224.1

Schrader, B. (1995). Infrared and Raman Spectroscopy. VCH, New York. DOI: https://doi.org/10.1002/9783527615438

C. V. Raman. Retrieved from http://en.wikipedia.org/wiki/C._V._Raman

Ferraro, J. R. (2002) Introductory Raman Spectroscopy. 2nd edtion. Academic Press. New York. ISBN: 9780122541056

McNay G., Eustace D., Smith W. E., Faulds K. & Graham, D. (2011). Surface-enhanced raman scattering (SERS) and surface-enhanced resonance raman scattering (SERRS): a review of applications. *Applied Spectroscopy*, 65, 825–837. DOI: https://doi.org/10.1366/11-06365.

Schlüker, S. (2011). Surface Enhanced Raman Spectroscopy: Analytical, Biophysical and Life Science Applications. Wiley-VCH. Weinheim. DOI: https://doi.org/10.1002/9783527632756

Stiles, P-L., Dieringer, J. A., Shah, N. C. & Van Duyne, R. P. (2008) Surface-enhanced Raman spectroscopy. *Annual Review of Analytical Chemistry*, 601–626. DOI: https://doi.org/10.1146/annurev.anchem.1.031207.112814

Shatz, G. C. & Van Duyne, R. P. (2002). Electromagnetic Mechanism of Surface-enhanced Spectroscopy. In J. M. Chalmers & P. R. Griffiths (Eds.), *Handbook of Vibrational Spectroscopy* (pp. 759–774), John Wiley & Sons. Chichester. DOI https://doi.org/10.1007/3-540-33567-6_2

Working Conditions in the Laboratory for Raman Microscopy

Steva Lević

Abstract

Raman spectroscopy/microscopy technique allows monitoring of various chemical processes in real time, and material characterization in terms of the distribution of individual chemical components via method known as Raman mapping. Equipping laboratory for this technique requires following: a stable supply of electricity, anti-vibration table for the instrument, constant temperature and controlled source of light. Prior to beginning the measurement it is necessary to select the laser of the appropriate wavelength and to calibrate the instrument.

The instruments that are used in Raman spectroscopy/microscopy allow monitoring of various chemical processes in real time, and material characterization in terms of the distribution of individual chemical components via method known as Raman mapping. To carry out complex analyzes of sample's surface it is necessary to possess a Raman microscope with motorized stand that allows automatic scrolling and analyzing previously given points on the sample. The resultant spectra form a "chemical map" of the sample.

How to cite this book chapter:
Lević, S. 2019. Working Conditions in the Laboratory for Raman Microscopy. In: Vucelić Radović, B., Lazić, D. and Nikšić, M. (eds.) *Application of Molecular Methods and Raman Microscopy/Spectroscopy in Agricultural Sciences and Food Technology*, Pp. 151–159. London: Ubiquity Press. DOI: https://doi.org/10.5334/bbj.j. License: CC-BY 4.0

In order to work in accordance with the requirements of the sample analysis, laboratory for Raman microscopy should meet several basic conditions. First of all, microscope needs a stable supply of electricity. It is necessary to eliminate sources of vibration that could adversely plug on the operation of the microscope, by placing it on the appropriate anti-vibration table (Figure 1). In addition, air-conditioning in the room is necessary in order to ensure conditions for the work of staff but also the stable temperature required for the smooth operation of the device.

In addition to the elimination of the effect of vibration, it is necessary to eliminate the influence of ambient light on the analysis in progress (Figure 2), which is achieved by shutting down the lights during excitation of the sample or by using microscope equipped with special curtain. Small table lamps that don't affect the operation of the device can be used.

1 Selection of laser of the appropriate wavelength i.e. choice of instrument

The first and basic condition for successful implementation of analysis by Raman microscopy is selection of the device i.e. laser. Older device types usually have only one laser that allows use of only one wavelength. Modern Raman

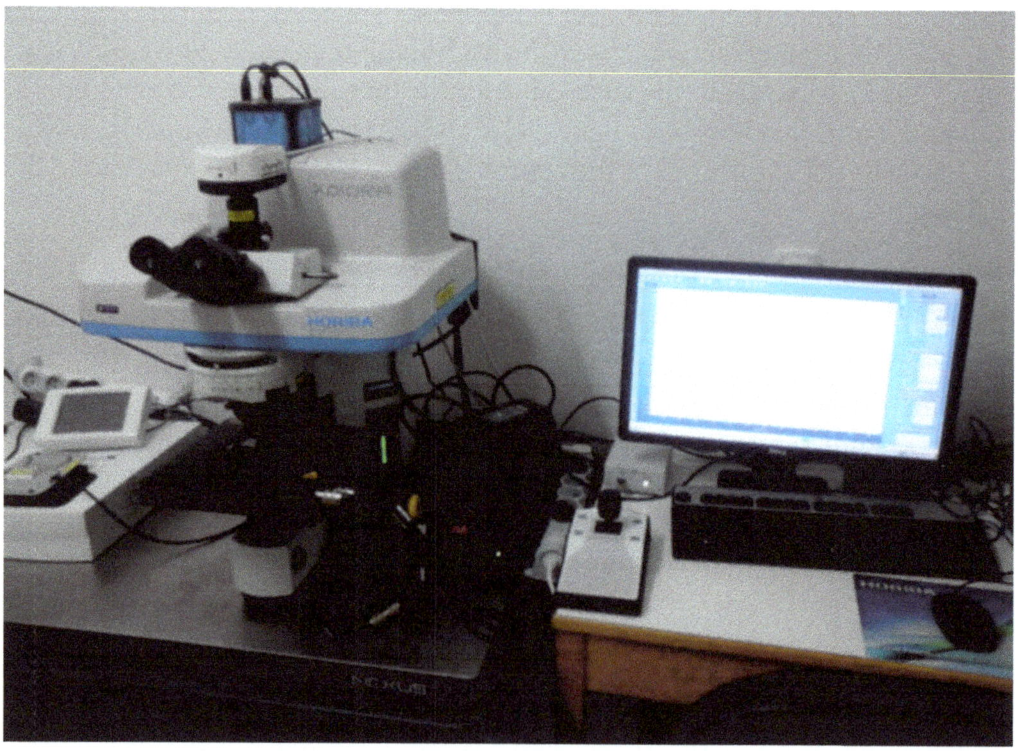

Figure 1: Raman microscope XploRA (Horiba Jobin Yvon) is set to the appropriate anti-vibration table.

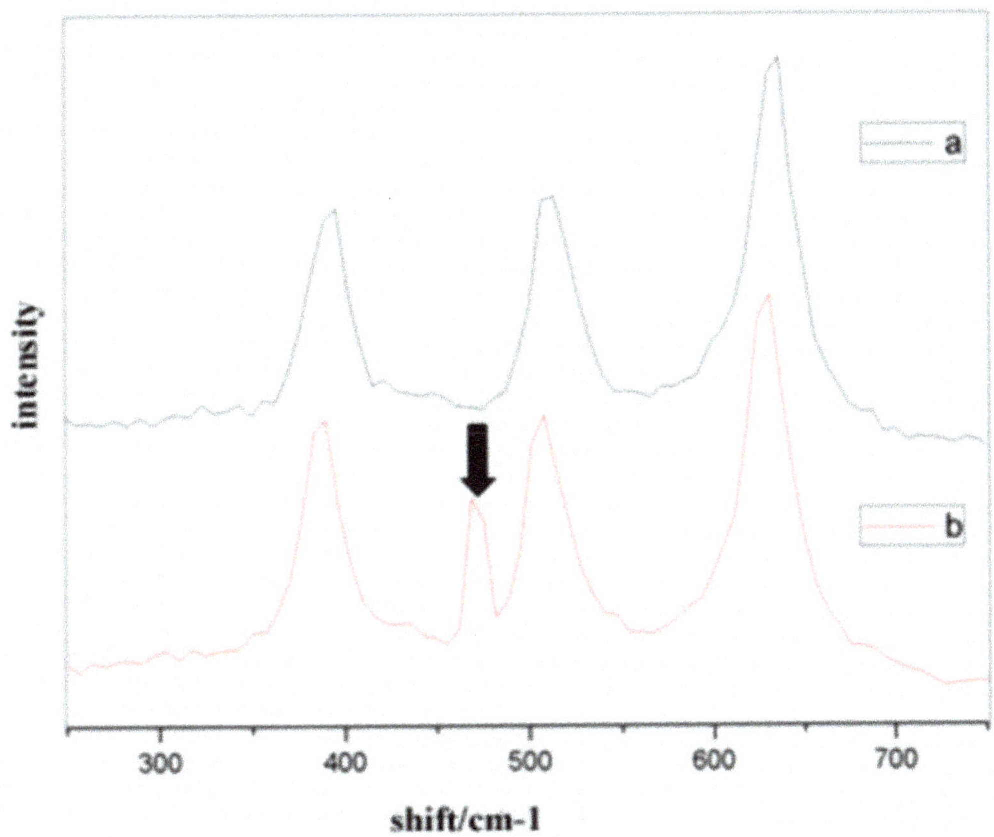

Figure 2: Influence of ambient light on the Raman spectrum of titanium dioxide. a) Recorded range without the influence of ambient light; b) includes the range of ambient light. The arrow marked signal from the ambient light. Laser 532 nm, acquisition time of 1s.

microscopes can be equipped with multiple lasers allowing quick switch from one source of excitation to another depending on the needs of the analysis or properties of analyzed material.

Selecting the appropriate laser primarily depends of the nature of tested material and its interaction with the laser. Specifically, by using certain lasers (e.g., 532 nm) fluorescence whose intensity exceeds the Raman signal (signal completely obscured) may occur. This phenomenon can be expressed so much that any technique for removing the fluorescence does not give any result. In such cases, lasers operating in higher wavelength range where the fluorescence is lesser or none (785 nm or 1064 nm) can be selected.

2 Calibration of the instrument

Prior to beginning the measurement it is necessary to calibrate the instrument. This is usually done by using pure substances of known positions of

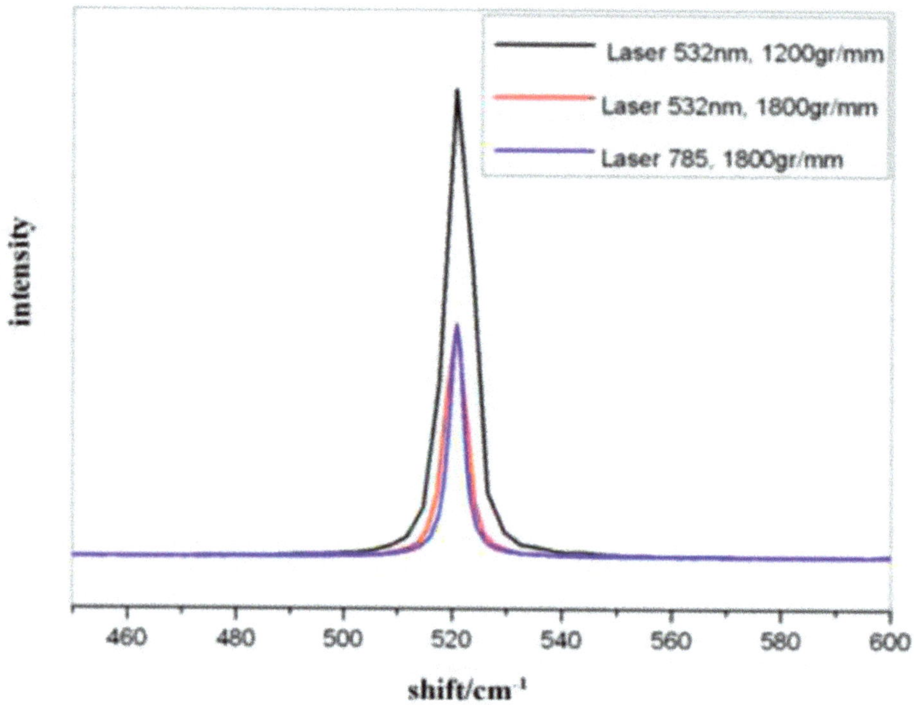

Figure 3: The result of using different lasers for calibration (calibration with silicon).

Raman signals in their spectra. Following substances can be used for calibration: silicon, sulfur, barium, etc. Example of calibration of Raman microscope XploRA (lasers at 532nm and 785nm) with silicon (peak at 520cm^{-1}) is shown in Figure 3.

Procedure of calibration depends on the type of device and the calibration procedure recommended by the manufacturer. Modern Raman microscopes have adequate indicator showing whether the calibration was made in accordance with the recommended procedure.

3 Procedure with samples

When it comes to working on Raman microscope, one should pay attention to some basic steps that should be taken before the analysis. The first and basic step is sample preparation prior to the measurement and the method of its conservation. It should be noted that the final form of the sample must be adopted to the dimensions of the microscope and its capabilities.

Generally, for Raman microscopy glass, quartz or gold (nanometer layer on the plate) microscopy slides are used. Unless otherwise specified, it is recommended to use quartz slides in order to eliminate as much as possible the influence of slide spectrum on the spectrum of the analyzed sample. Other substrates like a calcium fluoride are also available. Calcium fluoride could be used in the broad lasers' wavelength, with only on peak at around 320cm^{-1}. Also, in the recent years, SERS substrate based primarily on the silver nanoparticles are available, enabling Raman analysis of diluted samples.

If possible, one should avoid the use of plastic containers for holding samples during analysis because signals arising from the sample and from the container may overlap. In this case as well as when quartz plates are used, it is recommended to record spectrum of empty container prior to analysis, in order to avoid problems due to spectra overlapping.

3.1 *Analysis of solid samples*

Before the analysis it is necessary to properly apply a sample on the appropriate holder for recording. It is best to use quartz slides as this ensures minimum impact of the holder on Raman spectrum of the sample. When doing inverse Raman microscopy, the sample is analyzed from the bottom side and laser excitation of the sample is carried out from below. With such a configuration it is recommended to use a strip of two-sided-selfadhesive tape, first placing it onto a glass microscopic slide and then placing the sample on the other surface of the selfadhesive tape. Attention should be payed to possible negative impact of the strip material especially in the case of analyzing particles of smaller size and transparent samples because signals originating from the strip might appear in the sample spectrum.

3.2 *The analysis of liquid samples*

Most of the fluid samples may be analyzed immediately after application to the surface of the slide. This mode of recording is applicable for liquid samples that do not vary significantly (e.g. minimally evaporate) during analysis.

4 The effect of laser wavelength on the quality of Raman spectrum

The selection of the appropriate laser wavelength is certainly one of the most important tasks before starting the analysis (Figure 4).

When device is equipped with multiple lasers it is an easy procedure to change from one to the other laser. Before choosing the laser wavelength one should consult the literature from the field of investigation but also examine which laser is appropriate for particular sample.

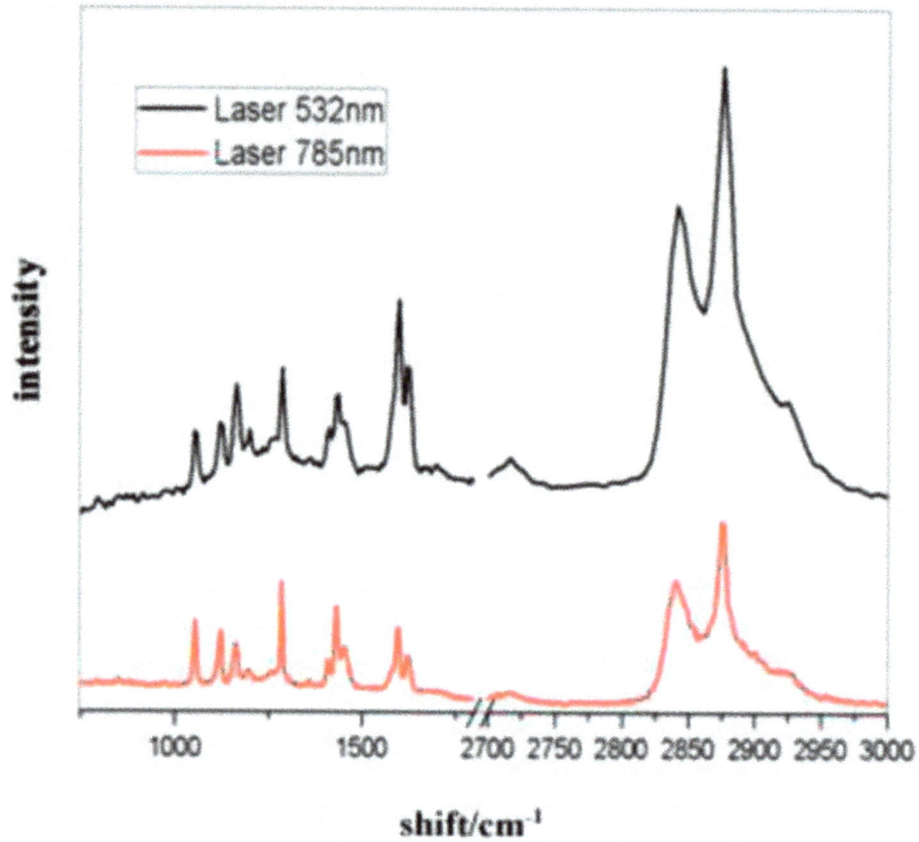

Figure 4: The effect of laser wavelength on Raman spectrum of carnauba wax.

However, the selection of the appropriate laser is limited to the available lasers that are installed. Lasers are part of larger optical system that includes adequate lenses, filters, spectrometer and detector.

5 The impact of laser power on the quality of Raman signal

The quality of the Raman spectra depends on the power of the applied laser. A general recommendation is to tend to obtain the maximum possible signal intensity. On the other hand it is necessary to take into account the stability of the samples that were exposed to the laser. In the Raman microscopy the laser is focused on a relatively small surface area which may lead to sample damage and can result in a signal which does not correspond to the actual spectrum of the sample. In extreme cases, too high intensity of the laser may completely disable the analysis due to the large sample damage. An example of such laser mediated damage of the sample is given in Figure 5.

It can be noticed that the laser greatly damaged the sample and therefore the analysis in these conditions is impossible. In order to prevent such a negative impact on the sample, the recommendation is to reduce the intensity of the

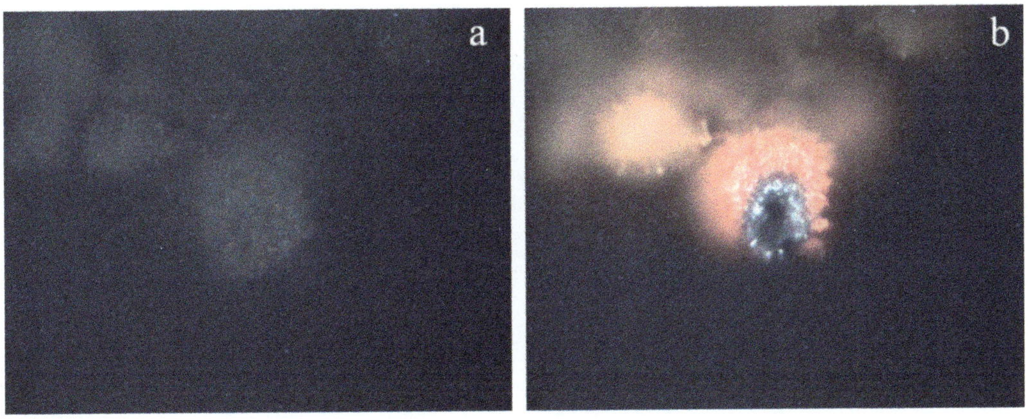

Figure 5: The impact of laser power on the sample: a) magnetite sample prior to analysis; b) after analyzing the sample of magnetite. Laser 532nm, filter 100%, acquisition time of 1s, the lens 50xLWD, laser power 25mW.

laser in combination with shortening duration of acquisition. The 25mW is enough, in the case of magnetite, to make considerable sample damage, while gradually reduction of laser power below 1mW (and combined with adequate acquisition time) usually provides satisfactory results. Other materials, like a previously mentioned titanium dioxide or carnauba wax could be analyzed using even greater laser power (e.g. 125mW), without visible sample damage.

Note: it is recommended to inspect the surface of the sample after the analysis in order to determine whether there has been any damage. In many cases it is possible to notice during recording that sample has been damaged. If severe damage of the sample occurs while recording in real time, the spectrum is rapidly shifting and usually at some point signal disappears (spectrum completely lost). This is the sign to stop recording, inspect the sample, confirm the damage and then change the parameters of the analysis in order to obtain a spectrum of adequate quality and avoid the sample damage. However, as previously mentioned, Raman microscopy analysis should aim at getting signal of high intensity in order to interpret the results correctly.

The impact of the laser intensity on Raman spectrum quality is shown in Fig. 6. Displayed spectra of titanium dioxide indicate that with any of the filters spectrum can be obtained, whereby the best signal is obtained at maximum intensity of the laser (filter 100%). Here also the rule applies that the use of laser at maximal intensity is dependent of the stability of the analyzed sample.

6 The effect of grooves density in the grating of spectrometer on quality of Raman signal

The number of lines in the grating of the spectrometer is set before recording. It is recommended to start with a smaller number of grooves and then if the analysis requires it increase this number. Higher number of grooves assures

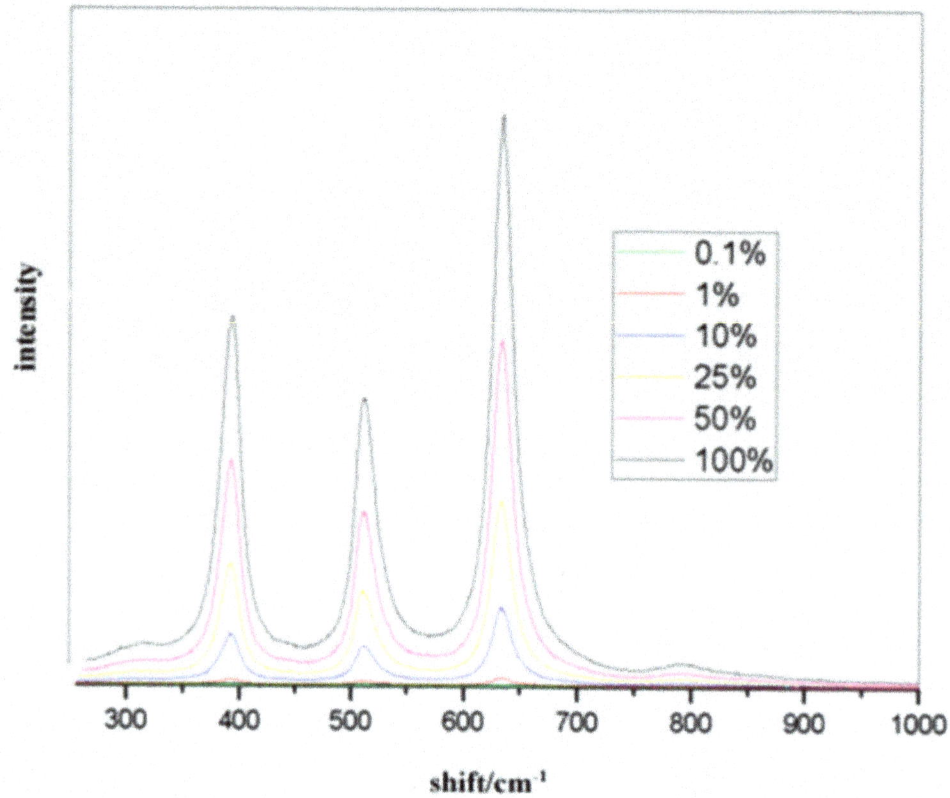

Figure 6: Effect of laser intensity on the quality of Raman spectrum of titanium dioxide. 532 nm laser, acquisition time of 1s, variable filters 0.1%, 1%, 10%, 25%, 50% and 100%.

better resolution and better separation of the peaks in the spectrum. It should be noted that using a higher number of grooves leads to the decrease in signal intensity as well as the increase in duration of recording. Signal intensity can be increased by increasing the duration of the recording.

Note: when recording involves a change in number of grating grooves, be sure to wait for the device to complete the procedure of changing the recording conditions to continue analysis under the new conditions. Program LabSpec 6 does not allow recording until the change of grooves number in the grating is not completed.

7 The influence of analysis duration on the quality of Raman spectrum

Duration of recording directly affects quality of the Raman signal. Generally, the longer is acquisition the obtained spectrum is more intense and therefore more reliable in terms of interpreting the results. Usually, the analysis of the sample begins with a shorter acquisition time which can then be extended in order to obtain a quality spectrum. On the other hand, too short acquisition time can lead to a drastic reduction in signal intensity and finally to a bad interpretation of the results.

Note: When selecting the time of acquisition one should pay attention to all possible phenomena. The first one is possible damage to the sample due to the long exposure to a laser. Another possible phenomenon is that the signal intensity is higher than the selected range of intensities, and the resulting spectrum is unusable. Therefore, the timing of acquisition is performed gradually, starting with a minimum of time that provides a signal and gradually increasing it until a maximal intensity of the spectrum is obtained.

Characterization of Microorganisms Using Raman Microscopy

Danka Radić

Abstract

Raman spectroscopy has recently gained popularity as an attractive approach for the biochemical characterization, rapid identification, and an accurate classification of a wide range of prokaryotes and eukaryotes organisms. In the case of eucariotes it is necessary to obtain higher number of Raman spectra in order to perform statistical analysis and to draw conclusions.

"*Raman spectroscopy (RS) is a powerful molecular fingerprinting technique which analyzes materials through the interaction of the material's molecules with an incident laser beam*" (Hanlon et al., 2000).

Vibrational spectroscopic technique, Raman spectroscopy (RS), has been used extensively to identify samples of different microorganisms by a careful investigation of the vibrating modes of the molecules in the microorganisms (Rösch et al. 2005). Raman spectroscopy has recently gained popularity as an attractive approach for the biochemical characterization, rapid identification, and an accurate classification of a wide range of prokaryotes and eukaryotes organisms (Hamasha 2011).

How to cite this book chapter:

Radić, D. 2019. Characterization of Microorganisms Using Raman Microscopy. In: Vucelić Radović, B., Lazić, D. and Nikšić, M. (eds.) *Application of Molecular Methods and Raman Microscopy/Spectroscopy in Agricultural Sciences and Food Technology*, Pp. 161–165. London: Ubiquity Press. DOI: https://doi.org/10.5334/bbj.k.

The Raman spectra of the microorganisms are superposition of spectra of the biochemical components inside the cells like e.g. protein, DNA, RNA, lipids, carbohydrates, water, as well as a few components with minor concentrations (Rösch et al. 2011).

Accordingly, the Raman spectra of two different species or strains show minor variations which originate from different chemical compositions due to variations in e.g. the cell wall (Rösch et al. 2011).

For the Raman spectroscopic characterization of eukaryotes like yeasts or fungi, different approaches are mandatory. It is not recommendable to use only one Raman spectra, in case of eukaryotes, because of the variations due to various organelles. An average of fifty spectra is necessary in order to perform statistical analysis and to draw conclusions. On this basis, it can be concluded, that the Raman spectroscopy can be used to identify yeasts or fungi (Stöckel et al. 2015).

For example, distribution of the width of the bands mirrors the different compounds and parts of the yeast cell (Figure 1a). Characteristic C = O stretch vibrations (1731–1765 cm^{-1}) represent the **lipid fraction**; mapping over the **amide I** region (1624–1687 cm^{-1}) produces bands arousing from the C= C lipid bonds. The **phenylenic** C = C Raman band (1567–1607 cm^{-1}) can be only seen in the periphery of the cells (*Rösch et al., 2005*).

During recording of a spectrum problems can occur. Fluorescence often appears when the examined material is complex and in color (Figure 1b), (*Jang and Akkus, 2013*). Another problem that might appear is burning of a sample (Figure 1c).

1 Preparation of the sample

1.1 Yeasts

The yeast cells were incubated at 28°C in a nutrient-rich YPD medium. Aliquots of the cell suspension were centrifuged (3000 rpm, 2min), washed three times with sterile water, and final suspended in the new aliquot of water (original suspension: water = 1:9). Spectra of the yeast (Fig. 1a).

Acquisition parameters

Recommendations for the spectra recording:

Laser wavelength: 532 nm, Greeting: 1200 gr/mm, Slit: 50 µm, Hole: 500 µm, Acquisition time: 20s, Range: 400–3200 cm^{-1}. Use a quartz plate!

1.2 Bacteria

The bacteria cells were incubated at 30°C–37°C in an appropriate nutrient-rich medium.

Sample preparationis the same as for the yeasts!

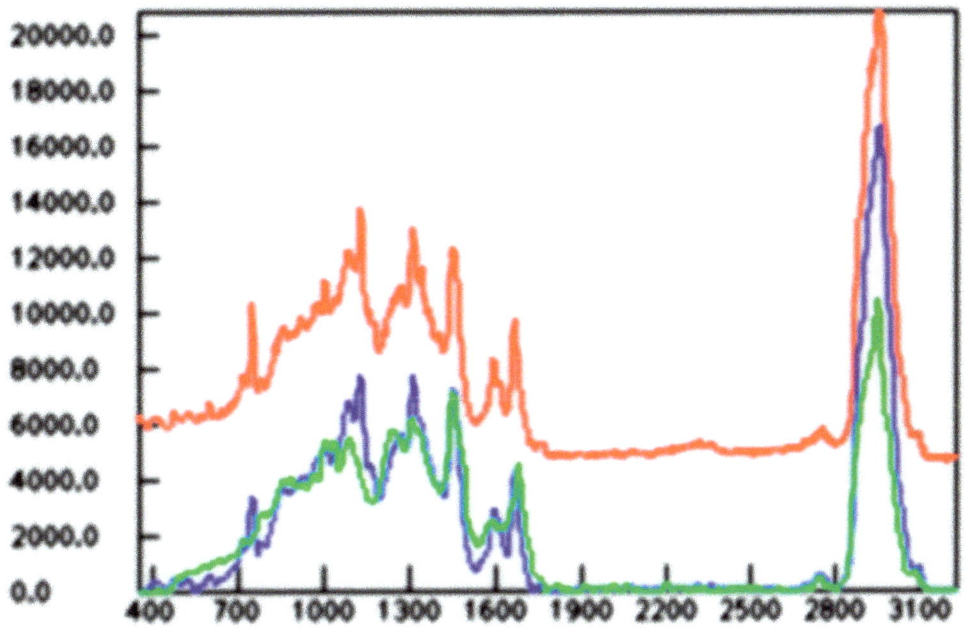

Figure 1a: Spectrum of the yeast. Different pronounced bands originated from lipids and proteins can be observed.

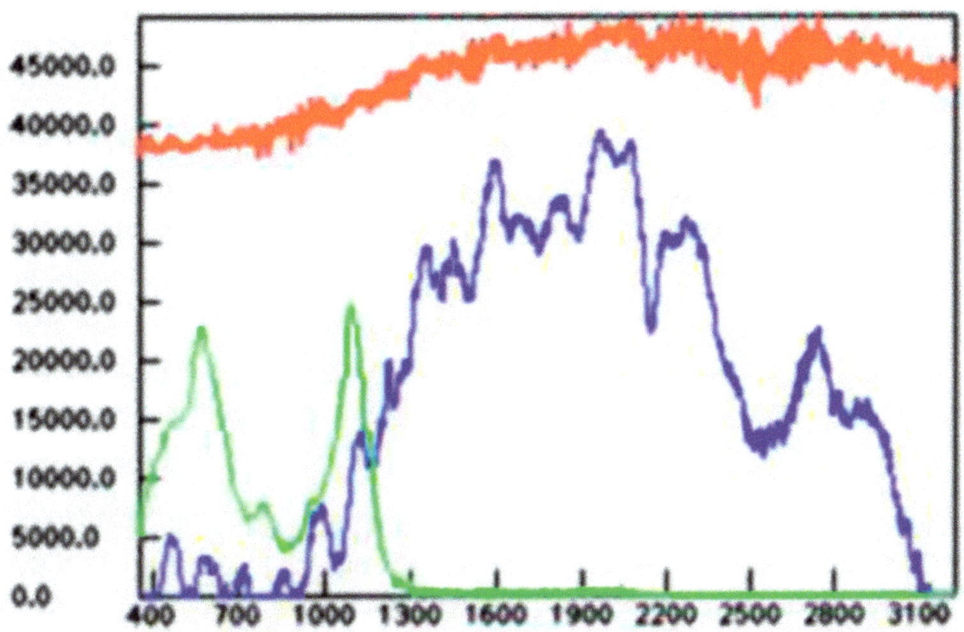

Figure 1b: Spectrum of fluorescence is marked by red label.

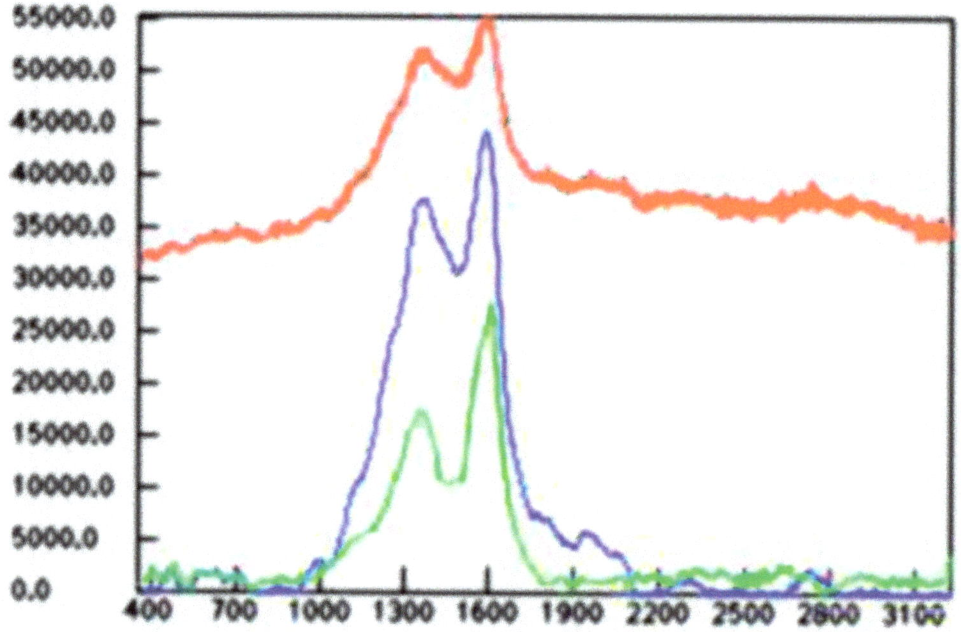

Figure 1c: Spectrum of burned cells is presented by red label.

Acquisition parameters

Recommendations for shooting:

Laser wavelength: 532 nm, Greeting: 600 gr/mm, Slit: 50 μm, Hole: 500 μm, Acquisition time: 30s, Range: 400–3200 cm^{-1}. Use a quartz plate!

In addition: All obtained spectra have to be processed using R program, which includes: spike removal, calibration, background removal, cutting, vector normalization, removing of the silent region.

2 References

Hamasha, K. M. (2011). *Raman spectroscopy for the microbiological characterization and identification of medically relevantbacteria* (Doctoral dissertation, Wayne State University)

Hanlon, E. B., Manohran, R., Koo, T. W., Shafer, T. W., Motz, J. T., Fitzmaurice, M., Kramer, J. R., Itkazan, I., Dasari, R. R., & Feld, M. S. (2000). Prospects for in vivo Raman spectroscopy. *Physics in Medicine & Biology*, 45(2), R1 59.

Rösch P., Harz, M., Peschke, K.D., Ronneberger, O., Burkhardt, H., & Popp, J. (2006). Identification of Single Eukaryotic Cells with Micro-Raman Spectroscopy. *Biopolymers*, 82, 312–316. DOI: https://doi.org/10.1002/bip.20449

Rösch, P., Harz, M., Schmitt, M., & Popp. J. (2005). Raman spectroscopic identification of single yeast cells. *Journal of Raman Spectroscopy,* 36, 377–379. DOI: https://doi.org/10.1002/jrs.1312

Rösch, P., Stöckel, S., Meisel, S., Münchberg, U., Kloß, S., Kusic, D., Schumacher, W., & Popp, J. (2011, November). *A Raman spectroscopic approach for the cultivation-free identification of microbes.* Paper presented at the SPIE 8311, Optical Sensors and Biophotonics III, Shanghai, China. DOI: https://doi.org/10.1117/12.901241

Stöckel, S., Kirchhoff, J., Neugebauer, U., Röscha, P., & Popp, J. (2015). The application of Raman spectroscopy for the detection and identification of microorganisms. *Journal of Raman Spectroscopy, Special Issue: International Year of Light,* 47(1) 89–109. DOI: https://doi.org/10.1002/jrs.4844

Yang, S. Akkus, O. (2013). Fluorescence Background Problem in Raman Spectroscopy: Is 1064 nm Excitation an Improvement of 785nm, Wasatch Photonics: Report 2013.

Polysaccharide Mushroom Extracts – Digging Into the Unknown

Jovana Vunduk

Abstract

Mushroom polysaccharide extracts are complex with many compounds that could block the signal which results in fluorescence that is covering some part or all the important data of the spectra. In other case the sample is emitting no signal at all. Mushroom polysaccharide extracts are shyly emerging in relation with Raman spectroscopy.

Polysaccharide mushroom extracts are complex systems obtained mainly by water extraction, and their prevailing compounds are polysaccharides, mainly glucans, which was confirmed by Raman studies of Moharram et al (2008). Depending on the mushroom species fruiting bodies could contain different proportions of different glucans.

Glucans are polymer compounds which consist of glucose units connected with α-(1,3), α-(1,6), β-(1,4) and β-(1,6) glycoside bonds (Synytsya and Novak,

---

How to cite this book chapter:

Vunduk, J. 2019. Polysaccharide Mushroom Extracts – Digging Into the Unknown. In: Vucelić Radović, B., Lazić, D. and Nikšić, M. (eds.) *Application of Molecular Methods and Raman Microscopy/Spectroscopy in Agricultural Sciences and Food Technology*, Pp. 167–170. London: Ubiquity Press. DOI: https://doi.org/10.5334/bbj.l.

2013). These complex substances are known to be responsible for biological effect of medicinal mushrooms, mainly as immunomodulators (Kozarski et al., 2015). Chemical composition of mushroom extracts greatly depends on the type of extraction. Thus characterization of different kind of extracts could elucidate the effect of the applied extraction procedure.

Theoretically Raman microspectroscopy provides an effective tool for the observation of biological material such as mushroom extracts. This method is

- undestructive
- does not acquire coloring agents to make the material visible
- there is no special preparation procedure for the sample
- measurement lasts from several seconds to several minutes

Literature data evidence for the different wavelengths that have been used in aim to characterize polysaccharide extracts; the main two are 850 and 1064 nm. Regions that belongs to polysaccharides are 1485–1464, 1363–1371, 1258–1267, 1118–1131, 1074–1084 and 1040–1048 cm^{-1} (Synytsya et al., 2009). Signals in the region between 750 and 950 cm^{-1} are from anomeric structures around glycoside bounds and they confirm α and β configuration (Mohaček-Grošev et al., 2001). One could expect complex spectra of the polysaccharide samples due to a very large number of possible vibrations (OH, C-H, C-C, C-C-O, C-O, C-O-C, C-O-H) (Lamrood et al., 2014).

1 Sample preparation

Raman spectroscopy does not require a special preparation procedure. The sample could be measured directly – in bulk, or dissolved in water. This technique is not invasive. The amounts that are required for the measurement are several mg.

2 Choosing the acquisition time

Acquisition time is the time during which the sample is exposed to the laser beam. This parameter is not fixed, so it should be chosen according to the real sample. One should make several spectra with several acquisition times and choose the time which gains the best spectra.

3 What is behind the expression „the good spectra“?

A good spectrum is the one whose characteristic peaks are clearly separated, without regions that are covered with fluorescence, since the fluorescence covers the distinctive lines and does not enable characterization. In order to obtain the statistical significance of the data approximately 50 good spectra should be collected.

4 Dead ends

Theoretically Raman spectroscopy is the perfect tool for the mushroom extracts observation. In reality the situation is less promising. The mushroom samples are complex with many compounds that could block the signal which results in fluorescence that is covering some part or all the important data of the spectra. In other case the sample is emitting no signal at all. The reason for such behavior is complexity of the samples. Pure β-glucan is clearly visible with distinctive lines, but the samples contain not just this type of glucans, but the phenolic compounds and proteins, too. These later might result in no signal.

5 Future perspective

Mushroom extracts are shyly emerging in relation with Raman spectroscopy which means that this topic is at its start (hopefully) or an inadequate technique. More attempts should be made, with a complex purification of the samples.

6 References

Kozarski M, Klaus A, Jakovljević D, Todorović N, Vunduk J, Petrović P, Nikšić M, Vrvić MM, van Griensven L. Antioxidants of edible mushrooms. Molecules, 2015, 20, 19489–19525.

Lamrood PY, Ralegankar SD, Harpale VM. Application of Raman spectroscopy for chemical characterization and protein conformation of *Agaricus bisporus* (Lange) Imabch (Agaricomycetidae) spores. International Journal of Bioassays, 2014, 3, 3318–3323.

Mohaček-Grošev V, Božac R, Pupples GJ. Vibrational spectroscopic characterization of wild growing mushrooms and toadstools. Spectrochemica acta Part A, 2001, 57, 2815–2829.

Moharram HA, Salama MF, Hussein AA. Characterization of Oyster mushroom mycelia as a food supplement. Australian journal of Basic and Applied Sciences, 2008, 2, 632–642.

Synytsya a, Mičkova K, Synytsya A, Jablonsky I, Spevaček J, Erban V, Kovarikova E, Čopikova J. Glucans from fruit bodies of cultivated mushrooms *Pleurotus ostreatus* and *Pleurotus eryngii*: Structure and potential prebiotic activity. Carbohydrate polymers, 2009, 76, 648–556.

Synytsya A, Novak M. Structural diversity of fungal glucans. Carbohydrate polymers, 2013, 92, 792–809.

Application of Raman Microscopy for Dairy Products Analysis

Aleksandar Nedeljković

Abstract

Raman microspectroscopy is applied for analysis of milk and dairy products.
The quality of spectra depends on type of the sample as well as on monitoring parameters. In the case of diluted samples problem can occure due to fluorescence of water. The choice of objective, duration of acquisition and temperature are essential for obtaining a good quality spectrum.

Raman spectroscopy, vibrational spectroscopic technique based on inelastic light scattering, provides qualitative and quantitative information about numerous types of samples and has therefore been applied in various research fields. Application of Raman spectroscopy in food science includes industrially oriented process and quality control, compositional analysis as well as more in-depth research utilization (examination of structure and structural changes of food components). Compared to another vibrational spectroscopic technique – IR absorption spectroscopy – Raman spectroscopy is more suitable for studying complex food systems, primarily due to the weak Raman scattering properties of water (Li-Chan 1996).

How to cite this book chapter:
Nedeljković, A. 2019. Application of Raman Microscopy for Dairy Products Analysis. In: Vucelić Radović, B., Lazić, D. and Nikšić, M. (eds.) *Application of Molecular Methods and Raman Microscopy/Spectroscopy in Agricultural Sciences and Food Technology*, Pp. 171–175. London: Ubiquity Press. DOI: https://doi.org/10.5334/bbj.m.

However, several questions arise during the course of Raman analysis. Therefore, in order to employ the full potential of this powerful technique and to acquire satisfactory spectra it is important to appropriately setup the Raman system by tuning several parameters (excitation wavelength i.e. laser, acquisition time, spectral range) and in addition to consider specific issues coming from the particular analyzed sample (Koca et al. 2010).

Raman spectroscopy is generally well-known for lack of sample preparation. As an illustration, for solid dairy products (cheeses, butter etc.) it is sufficient just to place the sample on the microscope slide. At the same time analysis of liquid dairy products (milk, yogurt etc.) requires employment of appropriate sample holder. Even though most of the Raman systems are equipped with specialized accessories for the analysis of liquid samples, where the cuvette is utilized, one can simply use custom-made sample container placed on microscope slide. For this purpose plastic cylinders are usually the best choice. Furthermore, usage of glass cuvette for analysis of milk has not shown best results because of the high turbidity of milk.

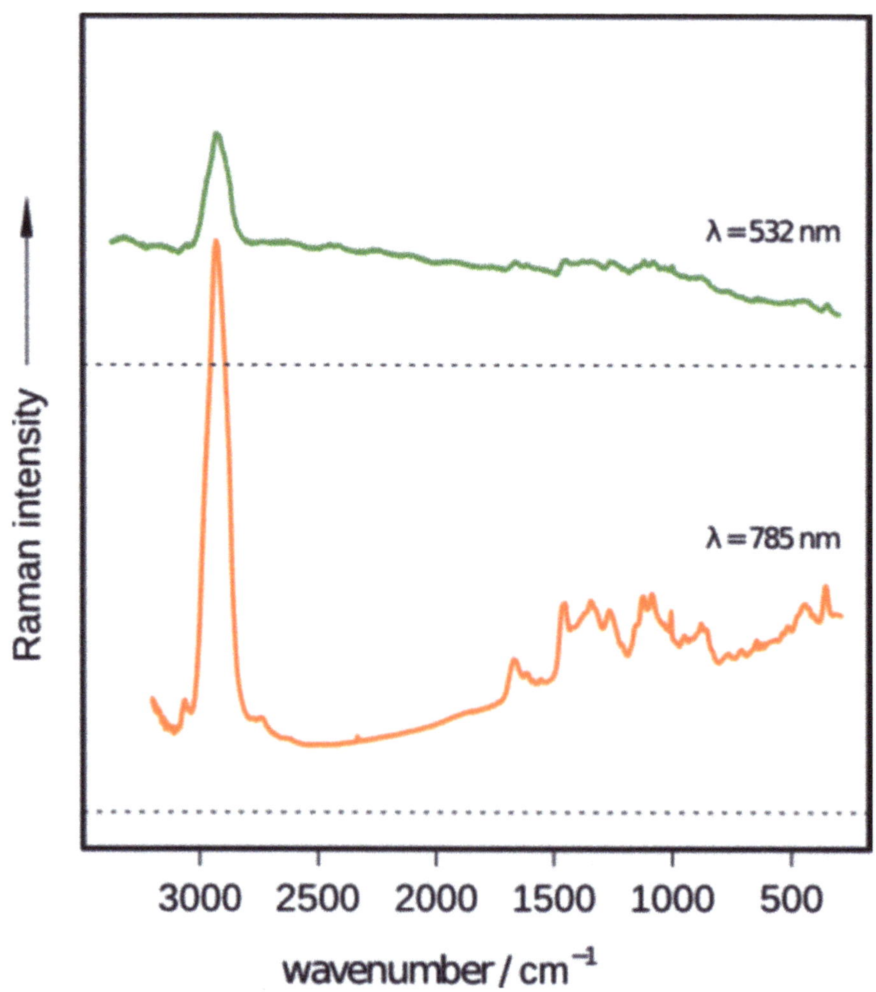

Figure 1: Fluorescence in analysis of milk samples.

Major drawback of Raman spectroscopy is the occurrence of fluorescence, and in the case of milk and dairy products it occurs with 532 nm lasers. According to Nedeljković et al. (2016) the spectral region 1800–800 cm^{-1} is important for milk fat analysis, which is almost completely overlapped by fluorescence by laser at 532nm (Figure 1). This drawback can be overcome by using lasers of a higher wavelength (785 nm, 1064 nm) (Figure 1.), at the expense of lower peak intensities. However, with the selection of slightly longer acquisition time, high-quality spectra can easily be acquired. Similarly, longer acquisition time (longer than 45 s) is also required for analysis of milk (Figure 2.). As can be seen, the short acquisition time provides only strong intensity of the CH_2 and CH_3 vibrations (around 2900 cm^{-1}), while increasing of the acquisition time enables clear visibility of peaks from milk sugars and proteins (Júnior et al. 2016) or from milk fat (Nedeljković et al. 2016). In contrast, with various "concentrated" milk systems (milk powders, cheese, kajmak etc.) where the water content is lower compared to milk, acquired spectra intensities are higher; hence acquisition time of 10 – 20 s is sufficient for satisfactory spectra.

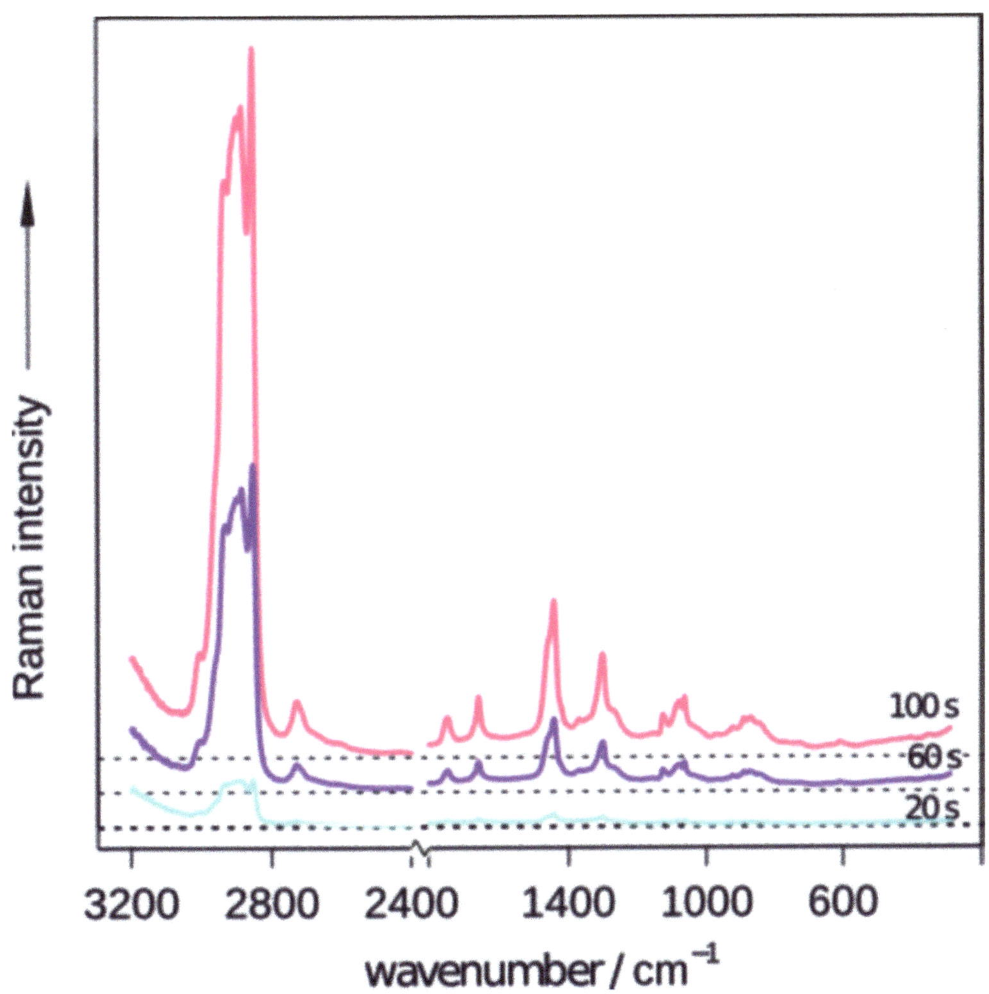

Figure 2: Influence of acquisition time on quality of milk spectra.

Another important feature in the Raman analysis of dairy products is the selection of objective which is used to focus the laser light to the sample. Generally if the specific task of the analysis is to examine the bulk of dairy product, and not to address the specific micro-structural issue, it is better to use the objective of lower magnification (x10, x5). The microscope can be used to get data successfully, but using small spot sizes can be very difficult for systems that can phase separate or contain domains of well-defined shape and refractive index – such as a lipid spheres i.e. milk fat globules. This can be addressed by taking multiple sample spot measurements on one sample or even mapping one of the samples to prove that domains are not a serious issue.

Finally, sample temperature during the spectra acquisition is quite important for Raman measurements (Abbas et al. 2009, Baeten et al. 2005). For that reason, consideration should be given to the selection of appropriate temperature and the selected temperature should be used throughout the given analysis. This is especially important for high-fat dairy products (cream, butter, high-fat cheeses) since the structural properties of the milk fat change considerably with the change of temperature and since the part of spectra in the "fingerprint" region corresponds to physical state of milk fat. Nevertheless, it was observed that holding the sample at room temperature for approximately 30 – 60 min prior to analysis would give adequate results. In the case of other temperatures, it is necessary to use temperature control system.

References

Abbas, O., Fernández Pierna, J. A., Codony, R., von Holst, C., & Baeten, V. (2009). Assessment of the discrimination of animal fat by FT-Raman spectroscopy. *Journal of Molecular Structure, 924–926*, 294–300. DOI: http://dx.doi.org/10.1016/j.molstruc.2009.01.027

Baeten, V., Fernández Pierna, J. A., Dardenne, P., Meurens, M., García-González, D. L., & Aparicio-Ruiz, R. (2005). Detection of the Presence of Hazelnut Oil in Olive Oil by FT-Raman and FT-MIR Spectroscopy. *Journal of Agricultural and Food Chemistry, 53*(16), 6201–6206. DOI: https://doi.org/10.1021/jf050595n

El-Abassy, R., Donfack, P., & Materny, A. (2009). Rapid Determination of Free Fatty Acid in Extra Virgin Olive Oil by Raman Spectroscopy and Multivariate Analysis. *Journal of the American Oil Chemists' Society, 86*(6), 507–511. DOI: https://doi.org/10.1007/s11746-009-1389-0

Júnior, P.H.R., de Sá Oliveira, K., de Almeida, C.E.R., De Oliveira, L.F.C., Stephani, R., da Silva Pinto, M., de Carvalho, A.F., Perrone, I.T. (2016). FT-Raman and chemometric tools for rapid determination of quality parameters in milk powder: Classification of samples for the presence of lactose and fraud detection by addition of maltodextrin. *Food Chemistry*, 196, 584–588. DOI: https://doi.org/10.1016/j.foodchem.2015.09.055

Koca, N., Kocaoglu-Vurma, N. A., Harper, W. J., & Rodriguez-Saona, L. E. (2010). Application of temperature-controlled attenuated total reflectance-mid-infrared (ATR-MIR) spectroscopy for rapid estimation of butter adulteration. *Food Chemistry, 121*(3), 778–782. DOI: http://dx.doi.org/10.1016/j.foodchem.2009.12.083

Li-Chan, E. C. Y. (1996). The applications of Raman spectroscopy in food science. *Trends in Food Science & Technology, 7*(11), 361–370. DOI: http://dx.doi.org/10.1016/S0924-2244(96)10037-6

Nedeljković, A., Rösch, P., Popp, J., Miočinović, J., Radovanović, M., Pudja, P. (2016). Raman Spectroscopy as a Rapid Tool for Quantitative Analysis of Butter Adulterated with Margarine. *Food Analytical Methods*, 9, 1315–1320. DOI: https://doi.org/10.1007/s12161-015-0317-1

Raman Microscopy in Plant Science, Carotenoids Detection in Fruit Material

Ilinka Pećinar

Abstract

Non-destructive nature of Raman analysis makes it exceptionally useful for various investigations of plant materials. It afforded the analysis of carotenoids in different fruits. The introduction of the NIR-FT-Raman technique led to many applications to green plant material by eliminating the problem of sample autofluorescence.

To gain a better understanding on structure, chemical composition and properties of plant cells, tissues and organs several microscopic, chemical and physical methods have been applied during the last years (Gierlinger & Schwanninger 2007). Raman spectroscopy with its various special techniques and methods has been applied to study plant biomass for about 30 years; such investigations have been performed at both macro- and micro-levels. Raman spectroscopy is in contrast to techniques like light microscopy, scanning electron microscopy (SEM), and transmission electron microscopy (TEM) which provide only morphological information of a material. Moreover, the non-destructive nature of Raman analysis along with none-to-minimal

How to cite this book chapter:

Pećinar, I. 2019. Raman Microscopy in Plant Science, Carotenoids Detection in Fruit Material. In: Vucelić Radović, B., Lazić, D. and Nikšić, M. (eds.) *Application of Molecular Methods and Raman Microscopy/Spectroscopy in Agricultural Sciences and Food Technology*, Pp. 177–186. London: Ubiquity Press. DOI: https://doi.org/10.5334/bbj.n.

requirement of sample preparation makes it exceptionally useful for various investigations. Raman spectroscopic applications on plants are very far ranging, going from investigations on structural polymers to metabolites to mineral substances (Gierlinger & Schwanninger 2007). In the field of plant science, Raman was first applied to study tracheid cells in the xylem of woody tissues in 1980s (*Atalla & Agarwal 1986*). Over the years, technological developments in the fields of filters, detectors, and lasers have made Raman instrumentation more suited to investigations of plant tissue (Agarwal 2014). Raman spectroscopy is an important method for investigating various plant tissues because it provides molecular level information on composition and structure of cellular components *in situ* without any staining or complicated sample preparation (i.e. cellulose and pectin: Atalla & Agarwal 1986, Gierlinger et al. 2010, carotenoids in tomato fruit: Qin et al. 2012, Baranska et al. 2006, starch, lipid and proteins in wheat grain: Manfait et al. 2004). It is a relatively specific spectroscopic technique that measures rocking, wagging, scissoring, and stretching fundamental vibrations of functional groups containing such bonds as C=C, C–C, C–O, C–H and O–H (Marquardta & Wold 2004). The major advantage of this technique is the capability to provide information about concentration, structure, and vibrational fingerprint of molecules within intact cells and tissues (Nikbakht et al. 2011). One of the main problems associated with the use of conventional Raman on plant materials is the very strong autofluorescence that is produced when phenolic compounds (i.e. lignin) are excited by visible light. In addition, the energies required to generate a Raman signal detectable above the autofluorescence, can cause heating and subsequent modification of the plant tissue. The introduction of the NIR-FT-Raman (1064 nm) technique led to many applications to green plant material by eliminating the problem of sample fluorescence (Agarwal 2014). For mapping and imaging of whole plant organs (seeds, fruits, leaves) the lateral resolution (about 10μm) of the NIR-FT technique is adequate, whereas for investigations on the lower hierarchical level of cells and cell walls higher resolution gained by a visible laser based system is needed. For investigation at the cellular level and it's compartments (i.e. carotenoids inside plant cells) resonance Raman spectroscopy could give promising results (Bhosale et al. 2014). Despite the fact that Raman scattering is an extremely weak by itself, when the energy of the scattered photon matches the energy of an electronic transition of the molecules, absorption and scattering of the chromophore are strongly increased (Meinhardt-Wollweber et al. 2018). This resonance effect may enhance the Raman spectrum by several orders of magnitude, where the molecules can be detected even at lower concentrations inside the sample. In large, complex molecular structures resonance selectivity helps to identify bands originating from vibrational modes of specific parts of the molecule, such as protein-cofactor complexes. In that way the target molecules could be recognized and enhanced above the others based on their resonance behavior.

1 Raman microspectroscopy and carotenoids detection in fruit material

Although carotenoids are minor components in plant material (lower than 0.1 mg per kg; Vitek et al. 2017), due to the specifics excitation of the Raman spectra in the visible wavelength excitation (532 nm), detection of carotenoids can be achieved by resonance Raman spectroscopy (Skoczowski & Troc, 2013; Zeise et al. 2018). Due to the high Raman activity of these compounds and the resonance effect resulting in a strong enhancement of the carotenoids bands, the resulting Raman spectroscopy has quite high potential for evaluating carotenoid biosynthesis (Vitek et al. 2017). Raman detection of carotenoid molecules allows strong and broad absorption bands for resonant excitation in the fluorescence-free wavelength at 532 nm, appropriate for sensitive detection of the molecule's highly specific Raman response (Bhosale et al. 2014). The Raman response of carotenoids is characterized by three strong high-frequency signals originating from C=C bond and C-C single bond stretches of the polyene chain and from methyl bonds (Koyama 1995). In this study investigation of different carotenoids was done on several fruit types of species such as: rose, nectarine, plum, pepper, maize and tomato. The aim of the study was recognising possible differences in carotenoids in different fruit types, eg. in a case of tomato fruit was observed carotenoids changes during fruit ripening, especially in pericarp and its parts.

2 Spectroscopic Measurements

The Raman spectra were recorded in the range 900–2000 cm^{-1} with a micro-Raman setup (HR LabRam inverse system, Jobin Yvon Horiba). Raman scattering was excited by a frequency-doubled Nd/YAG laser at a wavelength of 532 nm with a laser power incident on the sample of about 2 mW. The dispersive spectrometer has an entrance slit of 100 lm and a focal length of 800 mm and is equipped with a grating of 1200 lines mm^{-1}. Raman scattered light was detected by a CCD camera operating at 220 K. For the calibration procedure, titanium dioxide and 4-acetamidophenol (4AAP) were measured daily.

For the analysis of tomato carotenoids the following fruit stages were used: immmature green (20 DPA, days post antesis), mature green (39 DPA) and ripe fruit stage (52 DPA). On the fruit cross section three spots (pericarp, gel tissue and collumela) were measured for about ten times, and in total a minimum of 30 spectra were made from each fruit developmental stage. The total acquisition time per spectrum during the measurement was 1s, without filter corrections. These measurements were treated for baseline correction, normalization, cosmic spikes removal and intensity correction.

In Figure 1 it is clearly shown that carotenoids exhibit two strong Raman peaks in two separate spectral regions, 1100–1200 and 1400–1600 cm^{-1}, due

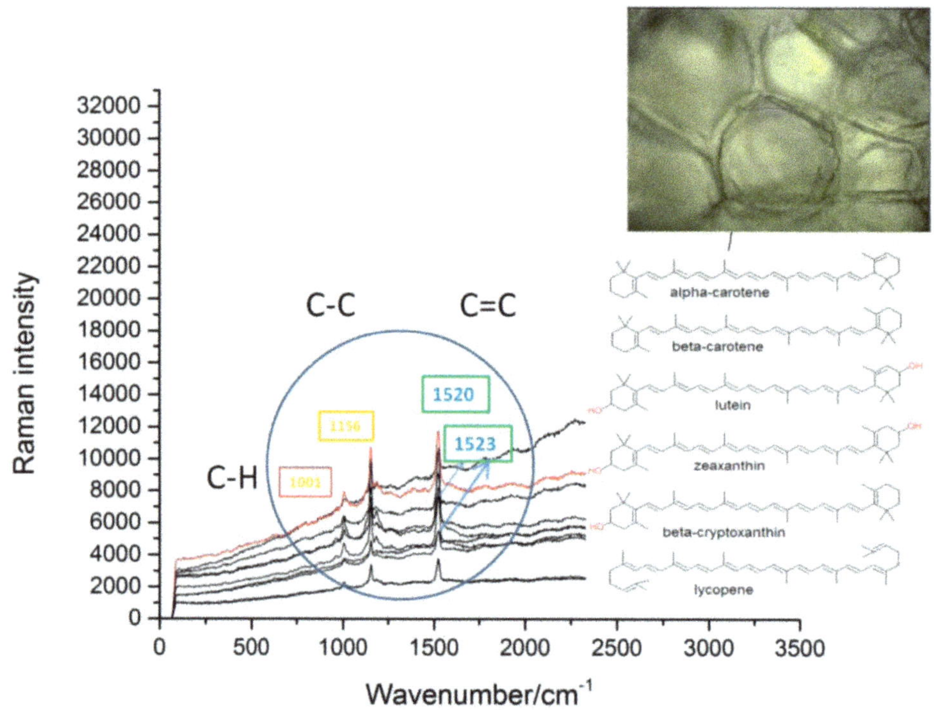

Figure 1: Raman spectrum of immature tomato fruit stage.

to the stretching vibrations of the C-C and C=C bonds in the polyene chain which composes the structure of carotenoids lycopene, β-carotene and α-carotene (Schulz & Baranska 2007). Based on the intensity of the spectra in tomato immature green stage as well as at mature green fruit stage, lutein can be detected at 1520 and 1523cm^{-1}. It can also be noted that no carotenoids were detected in the outer pericarp, exocarp, before the first phase of ripening while in mature green fruit stage peaks that indicate the presence of β-carotene and lycopene were observed.

Lycopene occurred first in gel issue at ripe fruit stage while before that signal was not observed. The intensity of lycopene signal was growing in gel tissue and in the whole pericarp during fruit ripening. In tomato fruit at ripe stage, the bands at 1.156 and 1.510 cm^{-1} are related to stretching vibrations of C-C and stretching vibrations of C=C of lycopene, respectively (Fig. 2). Moreover, the bands that appeared at 1.524 cm^{-1}, 1.157 cm^{-1} can be assigned to stretching C=C and stretching C-C of β-carotene, respectively (Schulz et al. 2006). Additionally, in-plane rocking mode of CH_3 groups attached to the polyene chain and coupled with C-C bands can be observed as a peak of medium intensity in the 1.000–1.020 cm^{-1} region which is obvious in Figure 2.

According to Figure 1 and considering the number of conjugated bonds in carotene and lycopene, it can be concluded that the peak observed at 1.520

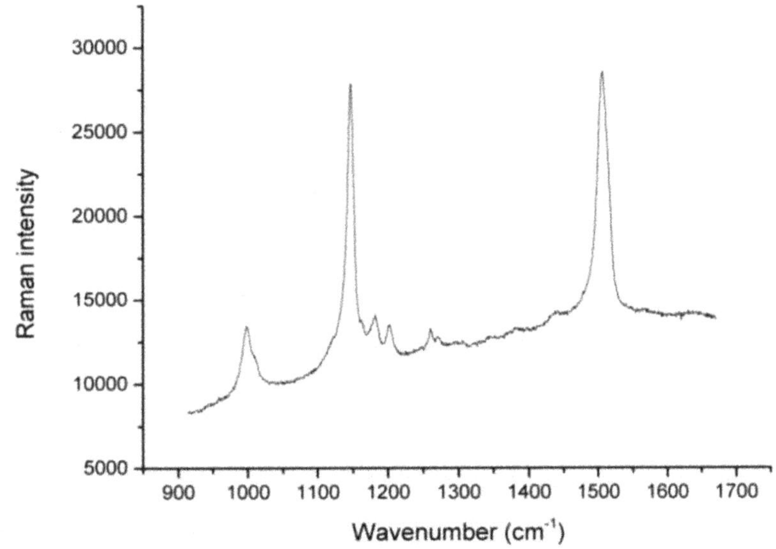

Figure 2: Raman spectrum of a red ripe tomato fruit at the region 900–1,600 cm^{-1}, having several peaks related to stretching vibrations of C-C and C=C groups.

cm^{-1} in Figure 2 implies the predominance of carotenes rather than lycopene. However, it can be seen that the band at 1.143 cm^{-1} (or 1.157 cm^{-1}) is asymmetric and there appear a three shoulders related to β-carotene or lycopene). The aim was to observe two separate spectral regions at better spectral resolution (grating 1200 lines mm-[1]) for better detection of carotenoids (Fig. 3). Detailed analysis of the above mentioned spectral regions was done with deconvolution of the spectra. This analysis was done by tool "peak analyzer" of Origine software using Voight function. With deconvolution of spectra some of parameters could describe different fruit carotenoids type, such as: peak area, relative intensity, position and full wide at half maximum (FWHM) of spectra. Some of parameters indicated on differentiations in tomato fruit regions related with ripening processes, especially peak relative intensity and peak area.

Similar Raman spectra for different kind of carotenoids were collected in other fruit types (Fig.4).

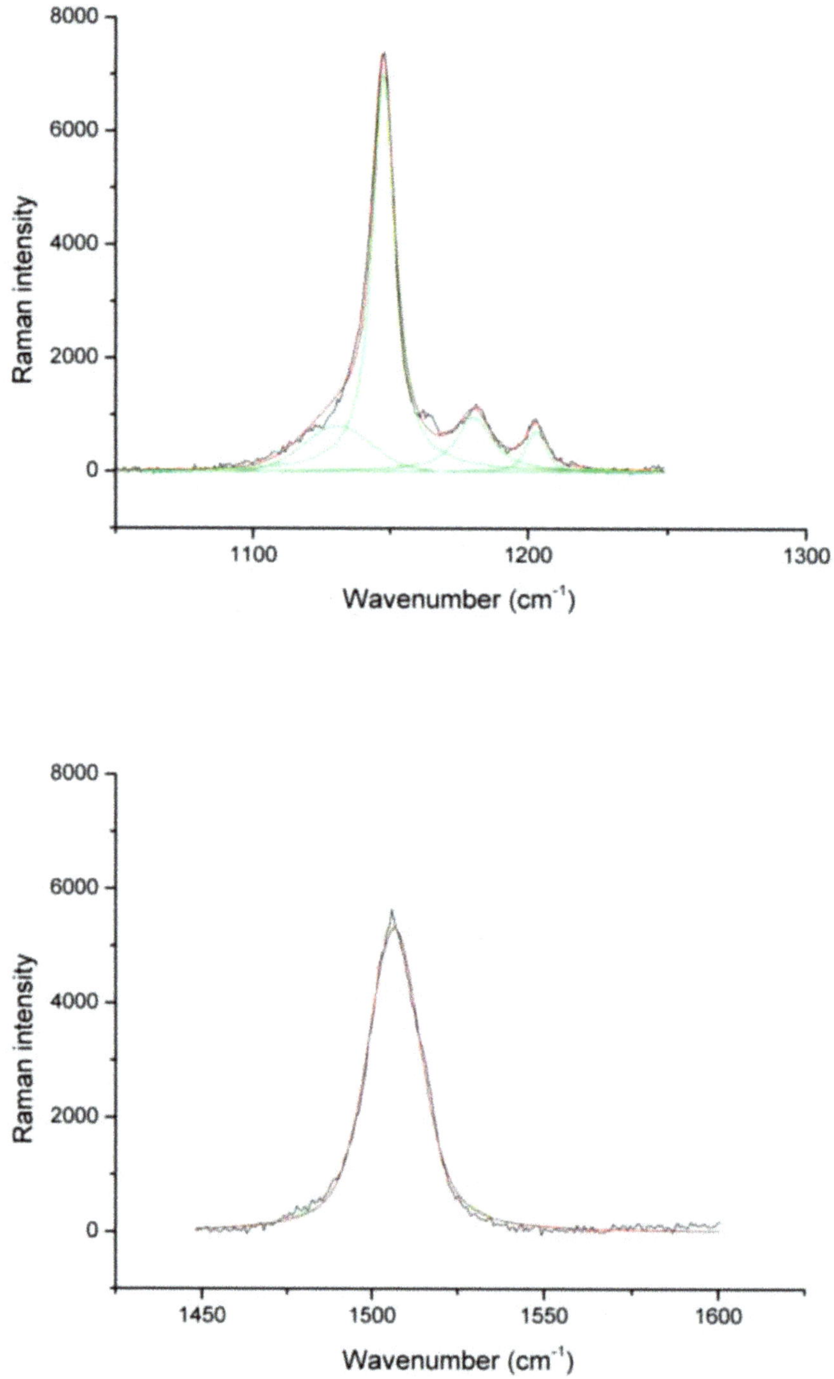

Figure 3: Deconvolution of two main carotenoids peaks in tomato fruit.

a)

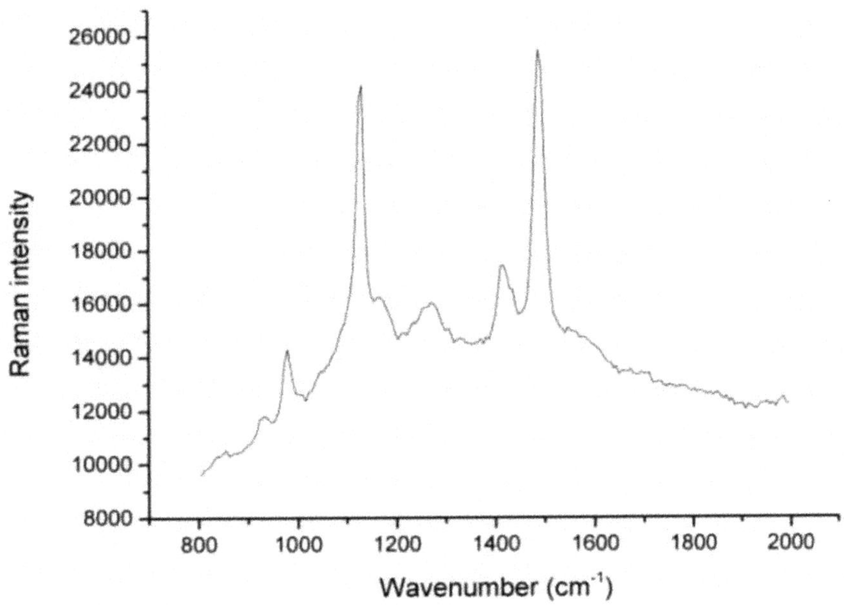

b)

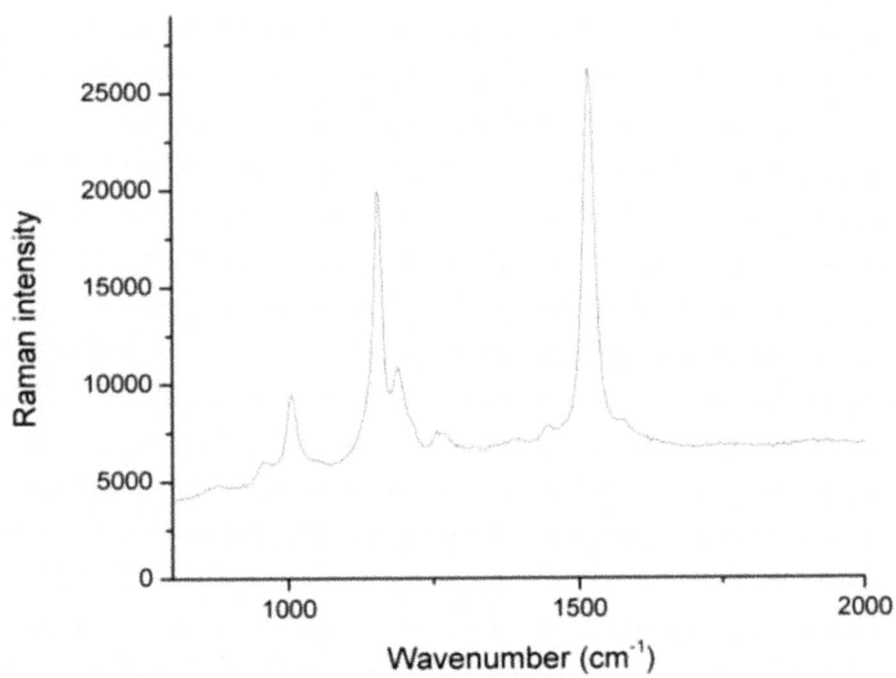

Figure 4: Raman spectra of carotenoids in other fruit types: a) *rose* hip, b) plum drupes, c) pepper berry, d) nectarine drupes, e) maize caryopsis.

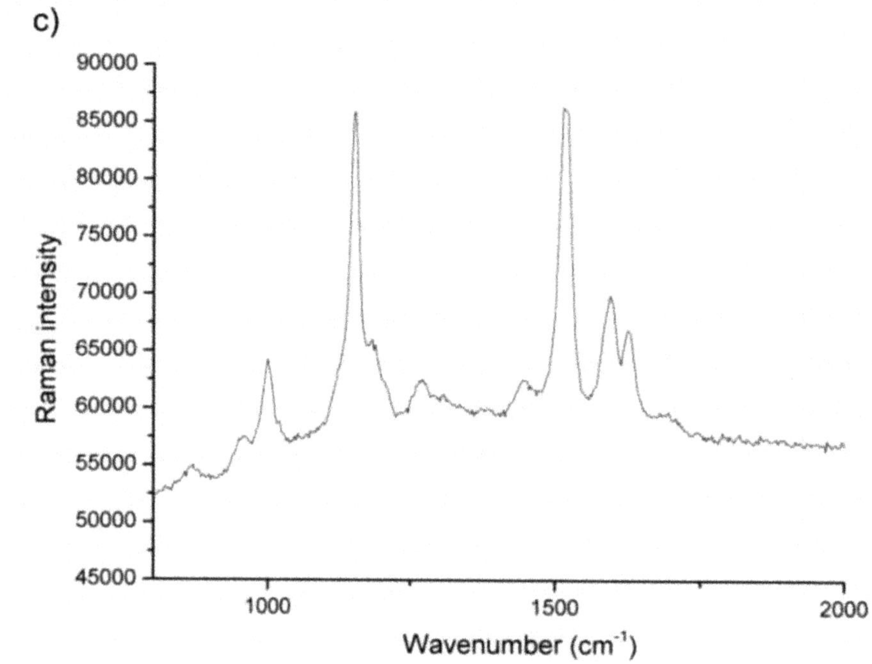

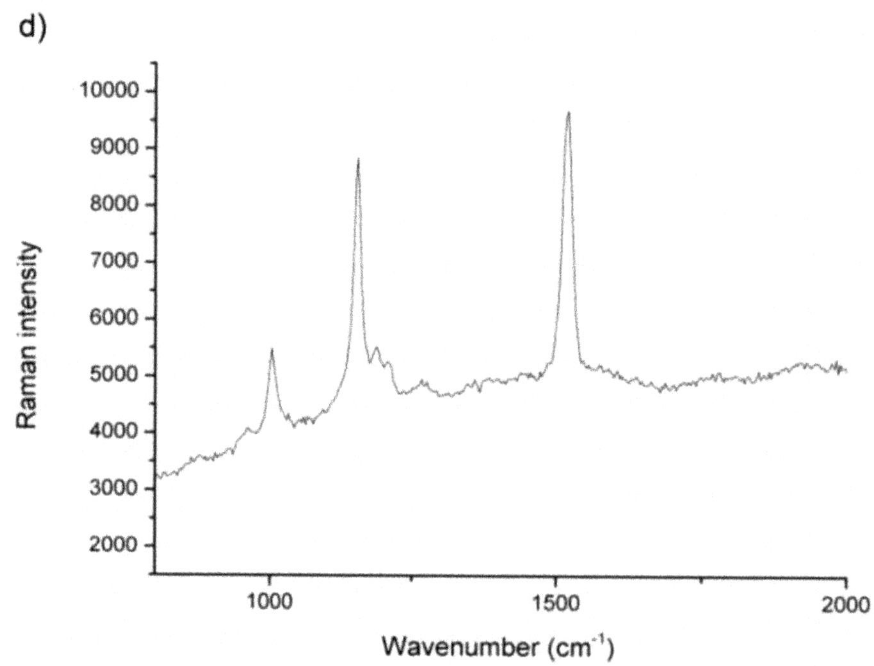

Figure 4: (*continued*).

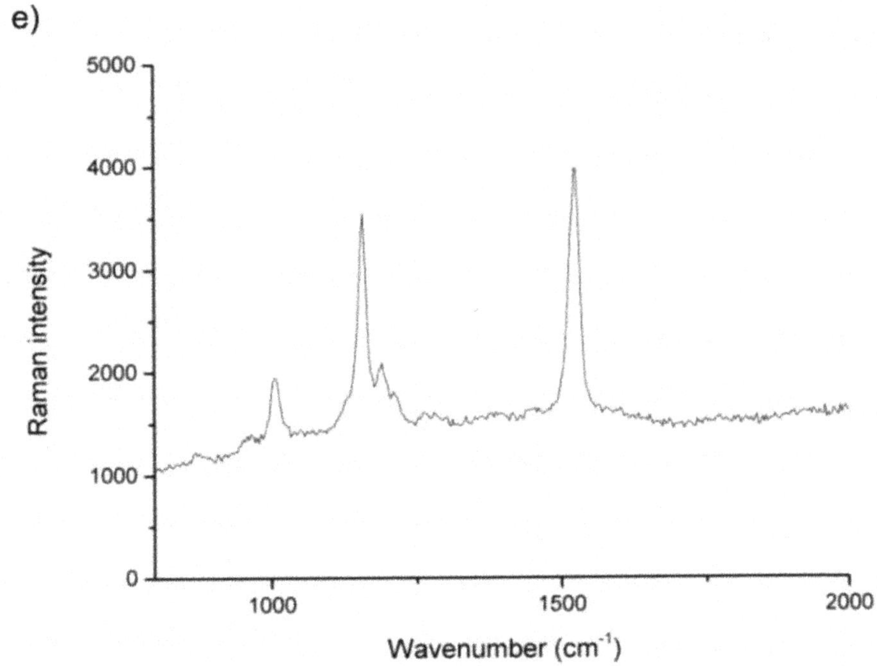

Figure 4: (*continued*).

3 References

Agarwal, U.P. (2014). 1064 nm FT- Raman spectroscopy for investigations of plant cell walls and other biomass materials. *Frontiers in Plant Science*. 5, 1–12. DOI: https://doi.org/10.3389/fpls.2014.00490

Atalla, R. H. & Agarwal, U. P. (1986). Recording Raman spectra from plant cell walls. *Journal of Raman Spectroscopy*, 17, 229–231. DOI: https://doi.org/10.1002/jrs.1250170213

Baranska, M, Schultze, W & Schulz, H. (2006). Determination of Lycopene and â-Carotene Content in Tomato Fruits and Related Products: Comparison of FT-Raman, ATR-IR, and NIR, *Spectroscopy. Analytical Chemistry,* 78, 8456–8461. DOI: https://doi.org/10.1021/ac061220j.

Bhosale, P., Ermakov, I. V., Ermakova, M. R., Gellermann, W. & Bernstein, P. S. (2004) Resonance Raman Quantification of Nutritionally Important Carotenoids in Fruits, Vegetables, and Their Juices in Comparison to High-Pressure Liquid Chromatography Analysis, *Journal of Agriculture Food Chemistry*, 52, 3281–3285. DOI: https://doi.org/10.1021/jf035345q

Gierlinger, N. & Schwanninger, M. (2007). The potential of Raman microscopy and Raman imaging in plant research. *Spectroscopy* 21, 69–89. http://dx.doi.org/10.1155/2007/498206.

Gierlinger, N., Luss S., König, C., Konnerth, J., Eder, M. & Fratzl, P. (2010). Cellulose microfibril orientation of *Picea abies* and its variability at the micron-level determined by Raman imaging. *Journal of Experimental Botany,* 61, 587–595. DOI: https://doi.org/10.1093/jxb/erp325.

Koyama, Y. (1995). Resonance Raman spectroscopy. *In Carotenoids*; Britton, G., Liaaen-Jenson, S., Pfander, H., Eds.; Birkhaeuser: Basel, Switzerland, Vol. 1B, pp 135–146.

Lang, P.L., Katon, J. E., & O'Keefe, J. F. (1986). The identification of fibers by infrared and Raman microspectroscopy. *Microchemical Journal,* 34, 319–331. DOI: https://doi.org/10.1016/0026-265X(86)90127-X

Manfait M., Piot, O., & Autran, J. C. (2004). Raman Mapping of Wheat Grain Kernels. Retrieved from http://www.horiba.com/fileadmin/uploads/Scientific/Documents/Raman/Bio04.pdf

Marquardta, B. J. & Wold, J. P. (2004). Raman Analysis of Fish: A Potential Method for Rapid Quality Screening. *Lebensm.-Wiss. u.-Technology*, 37, 1–8. DOI: https://doi.org/10.1016/S0023-6438(03)00114-2

Meinhardt-Wollweber, M., Suhr, C., Kniggendorf, A.-K. & Roth, B. (2018). Absorption and resonance Raman characteristics of β-carotene in water-ethanol mixtures, emulsion and hydrogel, *AIP Advances* 8:5 https://doi.org/10.1063/1.5025788

Nikbakht, A. M., Tavakkoli Hashjin, T., Malekfar, R. & Gobadian, B. (2011). Nondestructive Determination of Tomato Fruit Quality Parameters Using Raman Spectroscopy, *Journal of Agricultural Science and Technology,* 13, 517–526.

Qin, J, Chao, K. & Kim, MS. (2012). Nondestructive evaluation of internal maturity of tomatoes using spatially offset Raman spectroscopy. *Postharvest Biology and Technology,* 71, 21–31. https://doi.org/10.1016/j.postharvbio.2012.04.008

Schulz, H., Schütze, W. & Baranska, M. (2006). Fast determination of carotenoids in tomatoes and tomato products by Raman spectroscopy. *Acta Horticulture 712,* 901–906. DOI: https://doi.org/10.17660/ActaHortic.2006.712.118

Schulz, H. & Baranska, M. (2007). Identification and Quantification of Valuable Plant Substances by IR and Raman Spectroscopy. *Vibrational Spectroscopy, 43,* 13–25. http://dx.doi.org/10.1016/j.vibspec.2006.06.001

Skoczowski, A. & Troc, M. (2013). Isothermal Calorimetry and Raman Spectroscopy to Study Response of Plants to Abiotic and Biotic Stresses, *In Molecular Stress Physiology of Plants*, Editors: Gyana Ranjan Rout, Anath Bandhu Das, Springer Dordrecht Heidelberg New York London

Vítek, P., Novotná, K., Hodaňová, P., Rapantová, B., & Klem, K. (2017). Detection of herbicide effects on pigment composition and PSII photochemistry in *Helianthus annuus* by Raman spectroscopy and chlorophyll a fluorescence, *Spectrochimica Acta Part A: Molecular and Biomolecular Spectroscopy,* 170, 234–241. DOI: https://doi.org/10.1016/j.saa.2016.07.025

Zeise, I., Heiner, Z., Holz, S., Joester, M., Büttner, C. & Kneipp, J. (2018). Raman Imaging of Plant Cell Walls in Sections of *Cucumis sativus*, *Plants* 7, 7, DOI: https://doi.org/10.3390/plants7010007

Materials Characterization by Raman Microscopy

Steva Lević

Abstract

By using Raman microscopy it is possible to obtain various information such as chemical composition of material, its morphological and mechanical properties, distribution of specific compounds, structure of materials, etc. In order to localize the presence of specific compound in the complex mixtures such as in the case of solid capsules containing active compounds it is necessary to process its spectra.

In recent years Raman microscopy has attracted great attention as suitable technique for characterization of different solids and composites. By using Raman microscopy it is possible to obtain various information such as chemical composition of sample, its morphological properties, distribution of specific compounds, structure of materials, mechanical properties of sample, etc. The main advantages of Raman microscopy are reliability, generally easy data interpretation, non-destructive analysis of sample, possibilities for mapping of sample surface as well as deep profile analysis (Hollricher 2011, Fries and Steele 2011).

How to cite this book chapter:
Lević, S. 2019. Materials Characterization by Raman Microscopy. In: Vucelić Radović, B., Lazić, D. and Nikšić, M. (eds.) *Application of Molecular Methods and Raman Microscopy/Spectroscopy in Agricultural Sciences and Food Technology*, Pp. 187–192. London: Ubiquity Press. DOI: https://doi.org/10.5334/bbj.o.

The analysis of pure compounds is recommended before performing any chemical process or compounds mixing. Solid samples like crystals or powders can be analysed by simple deposition of material on the microscopic slide. Results could be obtained in the maters of seconds (Fig.1).

The Raman spectrum of ethyl vanillin (Fig. 1) exhibits several strong bands that are in agreement with chemical structure of this flavour compound. Usually, for material characterization the most important part of the spectrum is "fingerprint region", i.e. the part of spectrum with characteristic bands. For ethyl vanillin, "fingerprint region" could be defined in the spectral range 1000–1800cm^{-1}, where main bands are mainly associated with vibrations of the benzene ring.

Materials with crystal structure (like ethyl vanillin) can be relatively easily analyzed by Raman spectroscopy since the spectrum can be collected under short acquisition time and at maximum laser power.

When the spectra of pure compounds are known then it is possible to localize the presence of compounds in the complex mixtures. This is especially suitable in the case of formation of solid capsules that contain active compounds (e.g. encapsulated active compound). The efficiency of encapsulation process could be analysed using Raman microscopy in combination with mapping of particles' surface (Fig. 2).

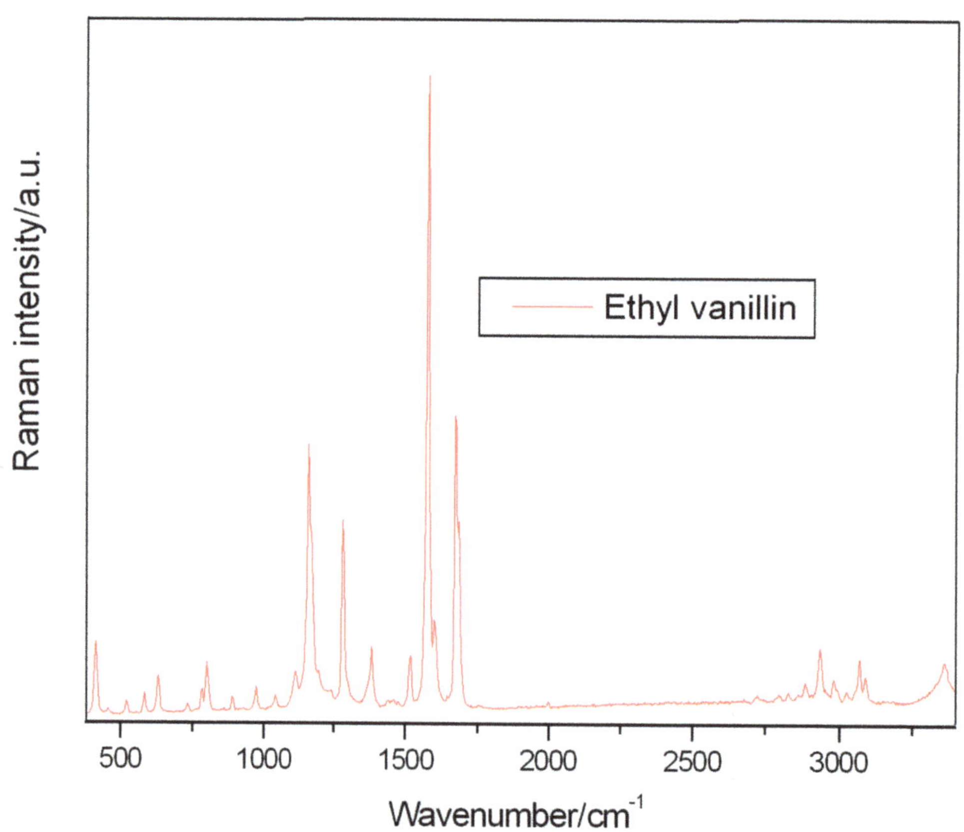

Figure 1: Raman spectrum of ethyl vanillin. Acquisition time 1s, laser 532 nm.

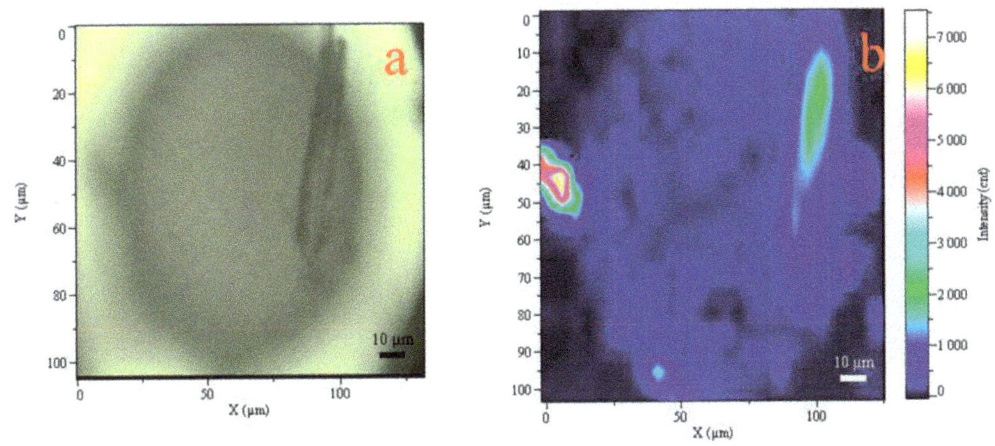

Figure 2: Raman mapping of carnauba wax microparticles and free ethyl vanillin that is not encapsulated into structure of carrier material (i.e. wax). Acquisition time 1s, step 1µm, laser 532 nm.

The free flavour is visible in the form of crystals on the particle surface. However, spherical shape of the particle prevents simultaneous identification of all crystals on the particle surface by light microscopy (Fig. 2a). Using Raman mapping it is possible to overcome this problem and as a result more visible ethyl vanillin crystals could be observed (Fig. 2b). The mapping spectra were collected using acquisition time of 1s and data were showed as maximum at $1576cm^{-1}$ (the most intensive band in the Raman spectrum of ethyl vanillin).

However, some analyses require longer acquisition time in order to ensure useful Raman spectra. In the Fig. 3 the two spectra of the same sample collected at two different acquisition times are showed. As can be seen, the longer acquisition time enables better visibility of bands at around $2900cm^{-1}$ (CH vibrations) which are important for cellulose characterisation (Agarwal et al. 2010).

Longer acquisition time is required when the intensity of main bands are not so strong and good for interpretation. In this case, Raman microscopy was found to be excellent for characterization of composite materials and successful determination of differences that occurred during sample preparation.

Characterization of natural polymers could be efficiently carried out by using Raman microscopy. As can be seen in the Fig. 4 Raman spectrum of sodium alginate exhibits several intensive bands that are linked to its structure and molecular properties. The use of Raman microscopy for such analyses usually requires optimisation of measuring method in order to provide useful data that can be further interpreted (see below).

Several key parameters must be adjusted before final sample analysis. The main parameter is selection of laser wavelength that will be used for analysis. Many materials exhibit strong fluorescence, especially when laser such as 532nm laser is used. Generally, in the case of high fluorescence it is recommended to use laser of higher wavelength (e.g. 785 nm or 1064 nm).

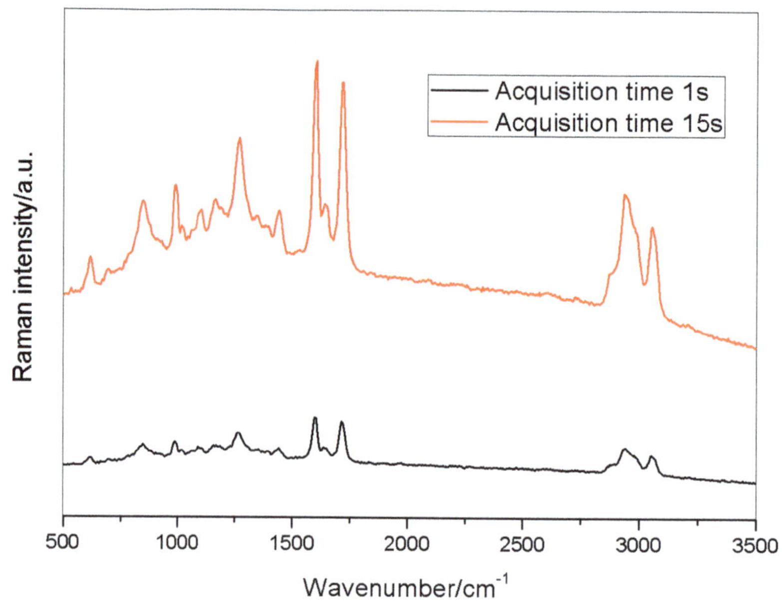

Figure 3: Influence of acquisition time on the spectra of cellulose based composite material. Laser 532nm.

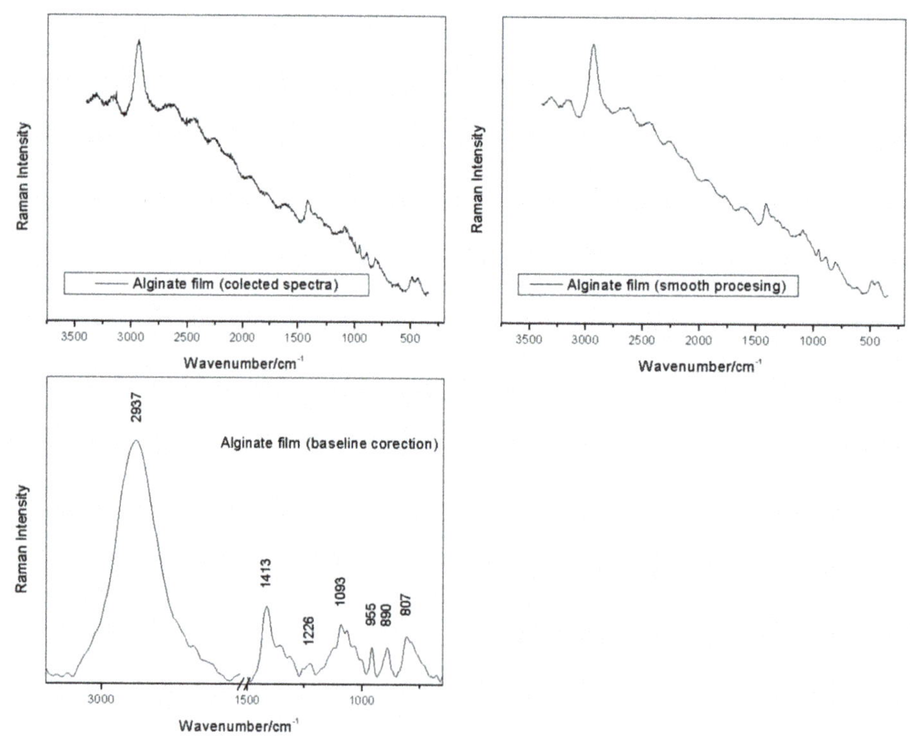

Figure 4: The post-analysis treatments of sodium alginate Raman spectra. Acquisition time 1s, laser 532 nm.

The data collected during Raman microscopy usually require some sort of treatment and interpretation before final conclusions. In the Fig. 4 the basic and processed Raman spectra of sodium alginate film are showed.

In the case of sodium alginate, the important bands are at 1413cm^{-1} (carboxylate group), while the bands between 1300–800cm^{-1} are related to the internal structure of alginate molecule and specific ratios of alginate subunits (Campos-Vallette et al. 2010).

The obtained spectrum of pure sodium alginate is subsequently processed by smoothing interpolation and baseline correction in order to eliminate as much as possible of data that do not belong to spectrum of alginate. Finally, processed spectrum obtained after baseline corrections and peaks identification provides data that could be used for further analyses.

The processing of obtained (basic) spectra can be divided in following operations:

1. Smoothing;
2. Baseline correction;
3. Intensity normalization;
4. Removing of suspicious points (i.e. peaks) form spectra.

These treatments are usually called a "pre-processing" of data and are required steps, especially when further analyses are necessary (e.g. statistical data analyses).

When an unknown compound is analysed, then the spectra of similar compounds could be found online (i.e. in the online bases of spectral data) and compared with analysed sample (http://sdbs.riodb.aist.go.jp/sdbs/cgi-bin/direct_frame_top.cgi). This could be useful for identification of compounds and results interpretation.

References

Agarwal, U.P., Richard S. Reiner, R.S., & Ralph, S.A. (2010). Cellulose I Crystallinity Determination Using FT-Raman Spectroscopy: Univariate and Multivariate Methods. *Cellulose*, 17, 721–733.

Campos-Vallette, M.M., Chandía, N.P., Ernesto Clavijo, E., Leal, D., Matsuhiro, B., Osorio-Román, I.O., & Torres, S. (2010). Characterization of Sodium Alginate and Its Block Fractions by Surface-Enhanced Raman Spectroscopy. *Journal of Raman Spectroscopy, 41*, 758–763.

Fries, M., & Steele, A. (2011). Raman Spectroscopy and Confocal Raman Imaging in Mineralogy and Petrography. In T. Dieing, O. Hollricher, & J. Toporski, J. (Eds). *Confocal Raman Microscopy* (pp. 111–135). Springer-Verlag, Berlin, Heidelberg. DOI: https://doi.org/10.1007/978-3-642-12522-5_6

Hollricher, O. (2011). Raman Instrumentation for Confocal Raman Microscopy. In T. Dieing, O. Hollricher, J. Toporski, J. (Eds), *Confocal Raman Microscopy* (pp. 43–60). Springer-Verlag, Berlin, Heidelberg. DOI: https://doi.org/10.1007/978-3-642-12522-5_3.

Spectral Database for Organic Compounds, AIST (SDBS). Retrieved from http://sdbs.riodb.aist.go.jp/sdbs/cgi-bin/direct_frame_top.cgi

Polarized Light Microscopy

Dragana Rančić

Abstract

Polarized microscopy and a variant of it polarized Raman spectroscopy are techniques that are used for analysis of anisotropic objects. It gives information on absorption, colour, structure, composition and other characteristics of different substances thus affording characterization and identification of variety of biological materials.

1 Polarized light microscopy

Polarized microscopy includes ilumination of the sample by polarized light. Normal, unpolarized, light (originated from both natural sunlight or most forms of artificial illumination) can be thought of as many sine waves, each oscillating at any one of an infinite number of planes around the central axis, while polarized light oscillates only in one plane. Unpolarized light can be transformed into polarized light, passing throught the special filter permitting the passage of light oscilating in only one plane (Fig. 1) (Collett 1993).

How to cite this book chapter:

Rančić, D. 2019. Polarized Light Microscopy. In: Vucelić Radović, B., Lazić, D. and Nikšić, M. (eds.) *Application of Molecular Methods and Raman Microscopy/Spectroscopy in Agricultural Sciences and Food Technology*, Pp. 193–198. London: Ubiquity Press. DOI: https://doi.org/10.5334/bbj.p.

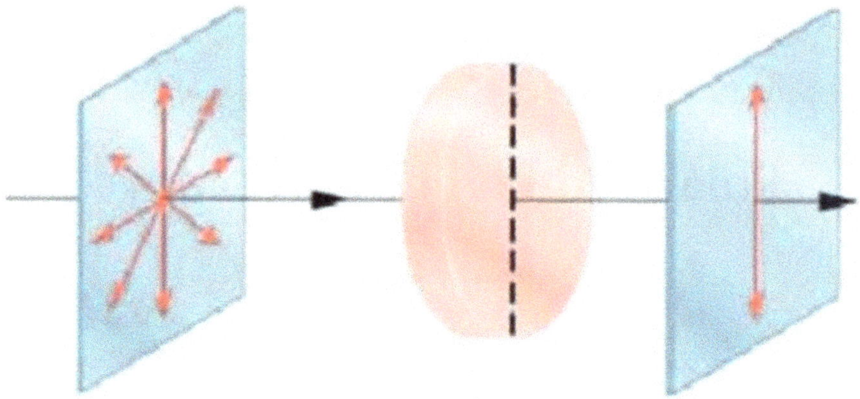

Figure 1: Polarizing filter allows passing only the light waves vibrating in a specific direction.

Microscopic examination using polarized light is especially useful for investigations of anisotropic objects. Isotropic materials, like unstressed glasses and cubic crystals, demonstrate the same optical properties in all directions, while anisotropic materials have optical properties that vary with the orientation of incident light. Around 90% of all solid substances are anisotropic. When these anisotropic materials are rotated, the observer may see brightness and color changes under polarized light that depend on the orientation of the material in the light path. This technique allows researchers to obtain information on color absorption, structure, composition, light refraction and other properties of different substances what can be used to characterize and identify various materials. Specimens having more than one index of refraction (eg. many organic as well as inorganic crystalline materials) produce birefringence effects, meaning that in field of view bright against the black background will appear. Views of some biological samples in polarization light can be similar to the darkfield technique, but polarized light is a contrast-enhancing technique that improves the quality of the image obtained with birefringent materials.

The polarized light microscope is equipped with two polarizing filters: first one (called polarizer) is positioned in the light path somewhere before the specimen, and the second one (called analyzer) is placed in the optical pathway after the objective rear aperture (Collett 1993). Polarized light is created by passing light through a polarizing filter, which transmits light in one direction only. An analyzer determines the amount and direction of light that illuminates a sample and image contrast arises from the interaction of polarized light with a birefringent specimen. When both the analyzer and polarizer are in the optical path, their permitted vibration directions are positioned at right angles to each other, so no light passing through the system and a dark field of view present in the eyepieces. Therefore, the image that is projected by the objective

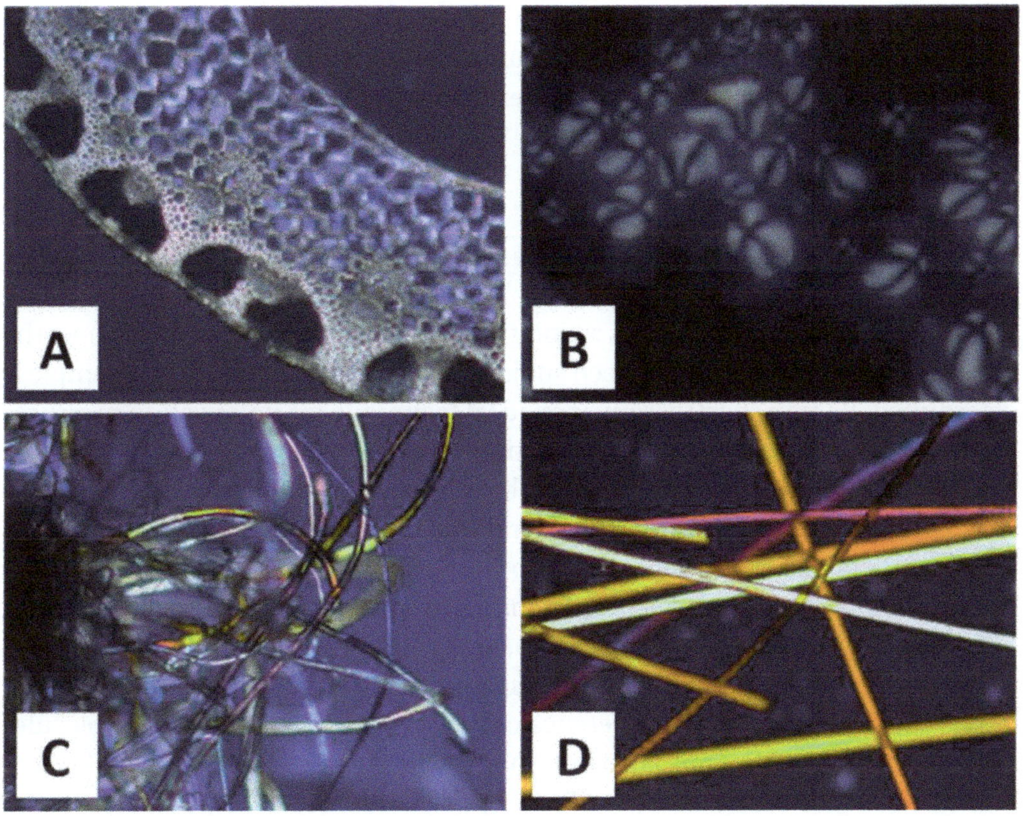

Figure 2: Some biological samples observed in polarizing light microscope: A- wheat stem cross section, B- potato starch, C- wool, D- human hair.

is mostly dark, except for bright specimen parts that are bierefringent or otherwise optically anisotropic (Fig 2).

Polarizing microscopy can be used with both, reflected or transmitted light. Transmitted light refers to the light diffused from below the specimen. This light is often passed through a condenser, which allows the viewer to see an enlarged contrasted image. Reflected light is especially useful for the study of opaque materials, such as metals, alloys, mineral oxides and sulfides.

Polarized light microscopy has a wide range of applications. The technique can be used both qualitatively and quantitatively in materials science, geology, chemistry, biology, metallurgy and medicine. Although polarized microscope is perhaps best known for its geological applications and is primarily used for the study of rocks and minerals, it can also be used to study many other materials including both natural and industrial minerals, composites such as cements, ceramics, mineral fibres, polymers, and crystalline. It is also useful in forensic examination. The various components in several biological specimens display beautifully in polarized light: different plant parts (Fig. 2A) especially fibres and wood, trichomes, pollen and starch grains (Fig. 2B) (Henry et al. 2011), bones, fish scales, wool (Fig. 2C) and hairs (Fig. 2D). It is also possible to observe cell division process and cell arrangement in tissues (Ivanov & Ignatov 2013), cellulose microfibrils and other details of plant cell wall (Leney 1981), silica body in plants (Dayanandan et al. 1983), numerous opaque and/or "thick" specimens and a wide variety of other specimens.

2 Polarized Raman scattering

Polarized Raman spectroscopy is a vibrational spectroscopic technique where the polarization of light which has been subjected to Raman scattering by a sample is determined. Raman spectroscopy is an ideal technique for polarization measurements because the lasers used for excitation emitte linearly polarized monochromatic light. But since Raman results can be misleading if one does not account for laser polarization, in conventional general purpose instruments the laser excitation beam is intentionally depolarized so as to give consistent results regardless of sample orientation. Advanced Raman instruments can be configured to provide, in addition to conventional Raman measurements, also polarized Raman measurements. In polarized Raman spectroscopy the polarized Raman scattering is observed as a result of the interference of the polarized light with the molecules vibrating. By using Raman polarization analyzer, the precise directional information about the differential polarizability of the molecules can be obtained, so polarized Raman spectroscopy probes provide insightful information such as molecular orientation and symmetries of the bond vibrations. Therefore, in addition to the general chemical identification which conventional Raman provides, these technique can improve the quality and quantity of the hyperspectral Raman dataset (Chiu et al. 2015) and it helps to reduce the depolarized auto-fluorescence backgrounds over the polarized Raman bands (Thomas et al. 1995).

This technique has been developed and usually applied for the chemical and physical analyses in material since, mainly in the aim to provide useful information relating to molecular composition and orientation in synthetic and natural polymer systems (Jarvis et al. 1980, Ward 1985, Bower 1972), but recently the role of polarization-resolved Raman spectroscopy in life science increased, opening new possibilities in monitoring biochemical composition, structures and symmetry of vibrations in biological tissues. It has already been used in the studying of various biomolecules including proteins, nucleic acids , crystals and fibers (Fanconi et al. 1969, Tsuboi et al. 1996, Tsuboi et al. 1997, Ko et al. 2006, Lim et al. 2011), in determining detailed architecture of viruses (George and Thomas 1999), evaluating three-dimensional collagen fibril orientation in tissues (Galvis et al 2013), Benevides et al 1997) or obtaining more chemical information from living cells (Chiu et al 2015). This technique also has applications in pharmaceutical science (Kiefer 2017) and has good diagnostic potential in medicine, for example in detection of cancer cells (Pachaiappan et al. 2017, Abramczyk et al. 2018).

3 References

Abramczyk, H., Brozek-Pluska, B., Kopec, M. (2018). Polarized Raman microscopy imaging: Capabilities and challenges for cancer research. Journal of Molecular Liquids, 259: 102–111

Benevides, J.M., Tsuboi, T., Bamford, J.H.K., Thomas, G.J. Jr. (1997). Polarized Raman spectroscopy of double-stranded RNA from bacteriophage φ6: local Raman tensors of base and backbone vibrations. Biophysical Journal, 72:2748–62

Bower, D. I. (1972). Investigation of molecular orientation distributions by polarized Raman scattering and polarized fluorescence. The Journal of Polymer Science (Polym. Phys.), 10: 2135–2153

Chiu, L, Palonpon, A. F., Smith, N.I., Kawata, S., Sodeoka, M., Fujita, K. (2015). Dual-polarization Raman spectral imaging to extract overlapping molecular fingerprints of living cells. Journal of Biophotonics, 8, 546–554.

Collett, E., (1993). Polarized Light: Fundamentals and Applications. Marcel Dekker, New York. ISBN-13: 978-0824787295

Dayanandan, P., Kaufman, P. B., & Franklin, C. I. (1983). Detection of Silica in Plants. American Journal of Botany, 70, 1079–1084.

Fanconi, B., Tomlinson, B., Nafie, L.A., Small, W., Peticolas, W.L. (1969). Polarized laser Raman studies of biological polymers. The Journal of Chemical Physics, 51(9):3993–4005.

Galvis, L., Dunlop, J.W.C., Duda, G., Fratzl, P., Masic, A. (2013) Polarized Raman Anisotropic Response of Collagen in Tendon: Towards 3D Orientation Mapping of Collagen in Tissues. PLoS ONE 8(5): e63518.

George, J., Thomas, Jr. (1999). Raman Spectroscopy of Protein and Nucleic Acid Assemblies. Annual Review of Biophysics and Biomolecular Structure, 28:1, 1–27

Henry, A. G., Brooks, A.S. , Piperno, D. R. (2011). Microfossils in calculus demonstrate consumption of plants and cooked foods in Neanderthal diets ((Shanidar III, Iraq; Spy I and II, Belgium). Proceedings of the National Academy of Science of the United States of America, 108, 486–491. DOI10.1073/pnas.1016868108

Ivanov, O. V. & Ignatov, M. S. (2013). 2D Digitization of Plant Cell Areolation by Polarized Light Microscopy. Cell and Tissue Biology, 7, 103–112.

Jarvis, D. A,. Hutchinson, I. J., Bower, D.I. and Ward, I. M. (1980). Characterization of biaxial orientation in poly(ethylene terephthalate) by means of refractive index measurements and Raman and infrared spectroscopies. Polymer, 21: 41–54.

Kiefer, J. (2017). Polarization-Resolved Raman Spectroscopy for Pharmaceutical Applications. American Pharmaceutical Review.

Ko, A. C.-T., Choo-Smith, L.-P., Hewko, M., Sowa, M. G., Dong, C. C. S., and Cleghorn, B. (2006). Optics Express Research, 14 (1): 203–215

Leney, L. (1981). A Technique for Measuring Fibril Angle Using Polarized Light, ISSN : 0043-7654

Lim, N. S. J., Hamed, Z., Yeow, C. H., Chan, C., and Huang, Z. (2011). The Journal of Biomedical Optics, 16(1): 017003.

Pachaiappan, R.; Prakasarao, A.; Singaravelu, G. (2017). Polarized Raman spectroscopic characterization of normal and oral cancer blood plasma. 100541F. 10.1117/12.2255600.

Thomas, G. J. Jr., Benevides, J. M. , Overman, S.A., Ueda, T., Ushizawa, K., Saitoh, M,Tsuboi, M. (1995). Polarized Raman spectra of oriented fibers of A DNA and B DNA: anisotropic and isotropic local Raman tensors of base and backbone vibrations. Biophysical journal, 68(3): 1073–1088.

Tsuboi, M, Overman, S.A., Thomas, G.J. Jr. (1996). Orientation of tryptophan 26 in coat protein subunits of the filamentous virus Ff by polarized Raman microspectroscopy. Biochemistry 35:10403–10

Tsuboi, M., Thomas, G.J. Jr. (1997). Raman scattering tensors in biological molecules and their assemblies. Applied Spectroscopy Reviews, 32:263–99

Ward, I. M. (1985). Determination of molecular orientation by spectroscopic techniques. In: Kaush H.H., Zachman H.G. (eds) Characterization of Polymers in the Solid State I: Part A: NMR and Other Spectroscopic Methods Part B: Mechanical Methods. Advances in Polymer Science, vol 66. Springer, Berlin, Heidelberg. ISBN 978-3-540-13779-5

Short Instructions for Raman Microscope Horiba Xplora

Dejan Lazić

When laser light hits the sample most of the light is reflected with no shift in the frequency. Part of photones transfer energy to the sample which irradiates it with the shifted frequency. Fluorescence occurs when the sample absorbes energy and releases it slowely. Instantaneous changes in frequency are Raman scattering. Shift in frequency, Raman effect correlates to the energy difference between ground and virtual state. Emmited light of lower energy than the energy of laser are called Stokes lines and ones of higher energy are called anti-Stokes lines. They are formed if electron was excited before light hit it. Concidering the fact that more molecules are in the ground state, it is expected to see more Stokes than anti Stokes lines. Antistokes lines could be interesting if Stokes lines are masked with fluoresence.

How to cite this book chapter:

Lazić, D. 2019. Short Instructions for Raman Microscope Horiba Xplora. In: Vucelić Radović, B., Lazić, D. and Nikšić, M. (eds.) *Application of Molecular Methods and Raman Microscopy/Spectroscopy in Agricultural Sciences and Food Technology*, Pp. 199–206. London: Ubiquity Press. DOI: https://doi.org/10.5334/bbj.q. License: CC-BY 4.0

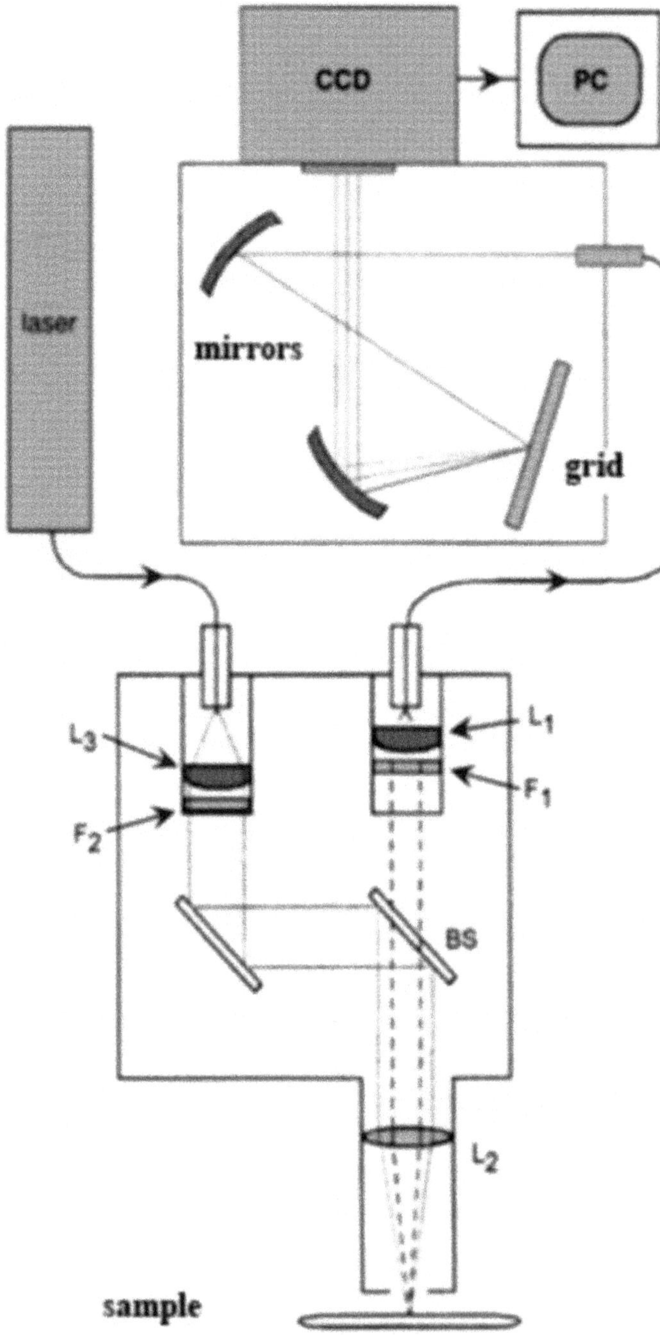

Figure 1: Construction of Raman microscope.

As Horiba Xplora instrument is the only Raman microscope at the Faculty of Agriculture further short instructions are aimed to facilitate its use by newly trained staff and PhD students in their reseach.

Turn on by pressing a switch on the extension cord. Switch on the computer, start LabSpec 6, icon on the middle of the screen.

Chech if lasers are powered on.

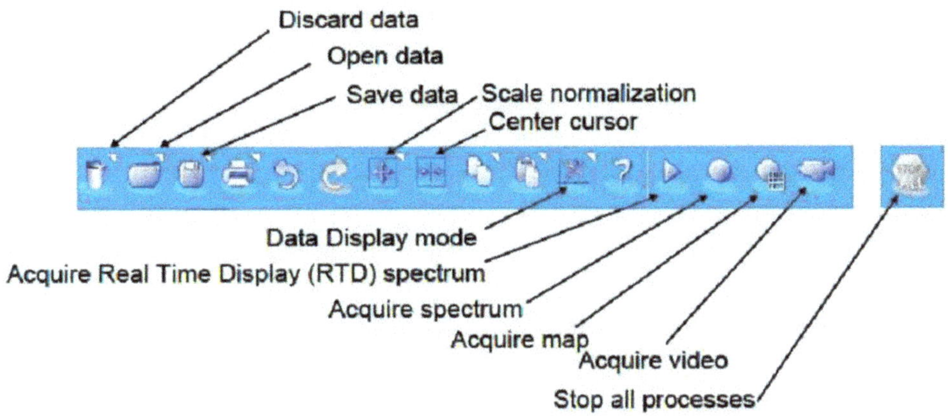

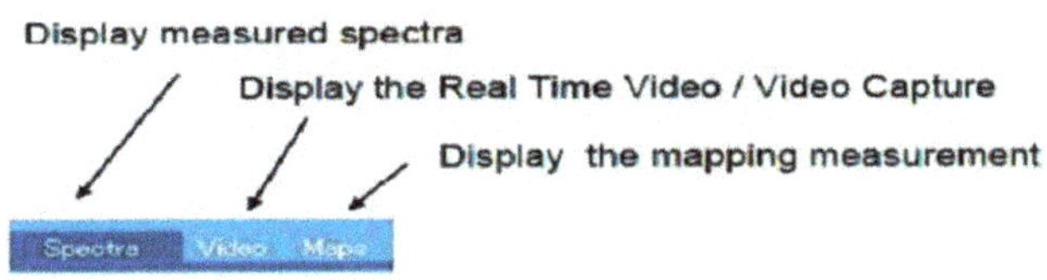

Control panel

The Control Panel on the right hand side includes all functionality required in LabSpec 6 to acquire, process, analyze and display data. The individual tabs of the Control Panel are organized as follows:

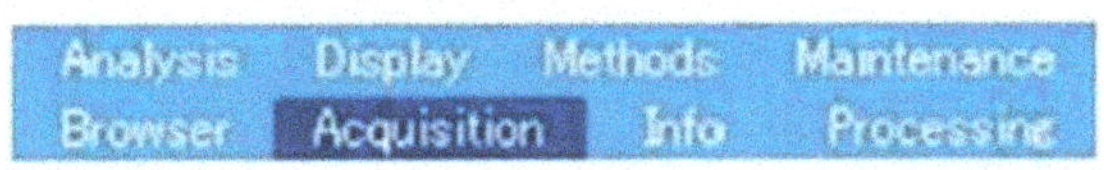

Browser: view a list of all data opened in LabSpec6.
Acquisition: set up all acquisition parameters including hardware settings.
Info: view the information shell for each data file
Processing: data processing functions to modify raw data (including smoothing, baseline correction and math functions)
Analysis: data analysis functions to obtain information from the data **Display:** configure the display of spectra / video / mapping window
Methods: create customized multi-step one click sequences, data acquisition and analysis.
Maintenance: system calibration, Auto Calibration and Auto Alignment

Autocalibration

If you see a red AC at the bothom of the screen autocalibration is required.

Figure 2: Basic commands on LabSpec 6.

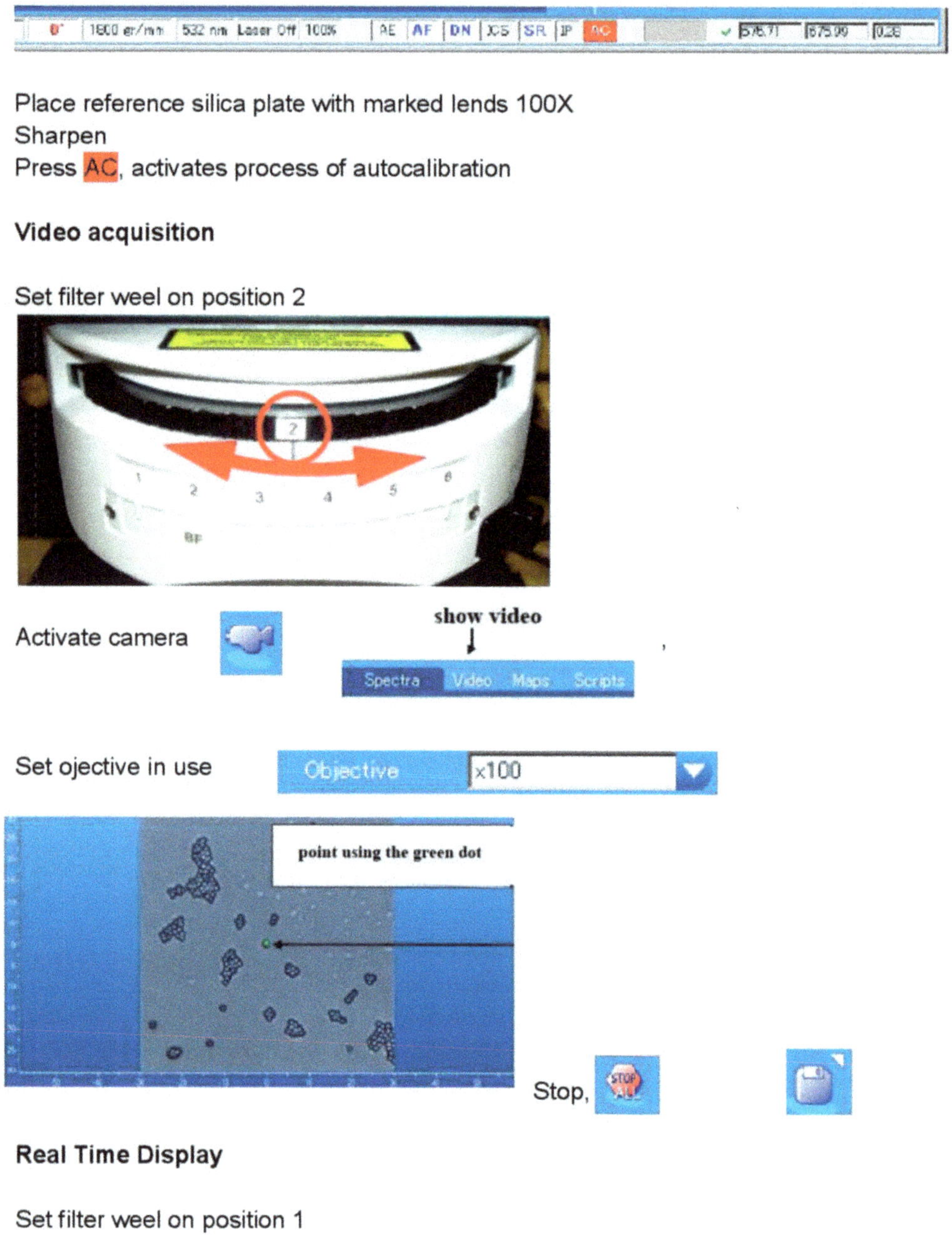

Figure 3: Video acquisition.

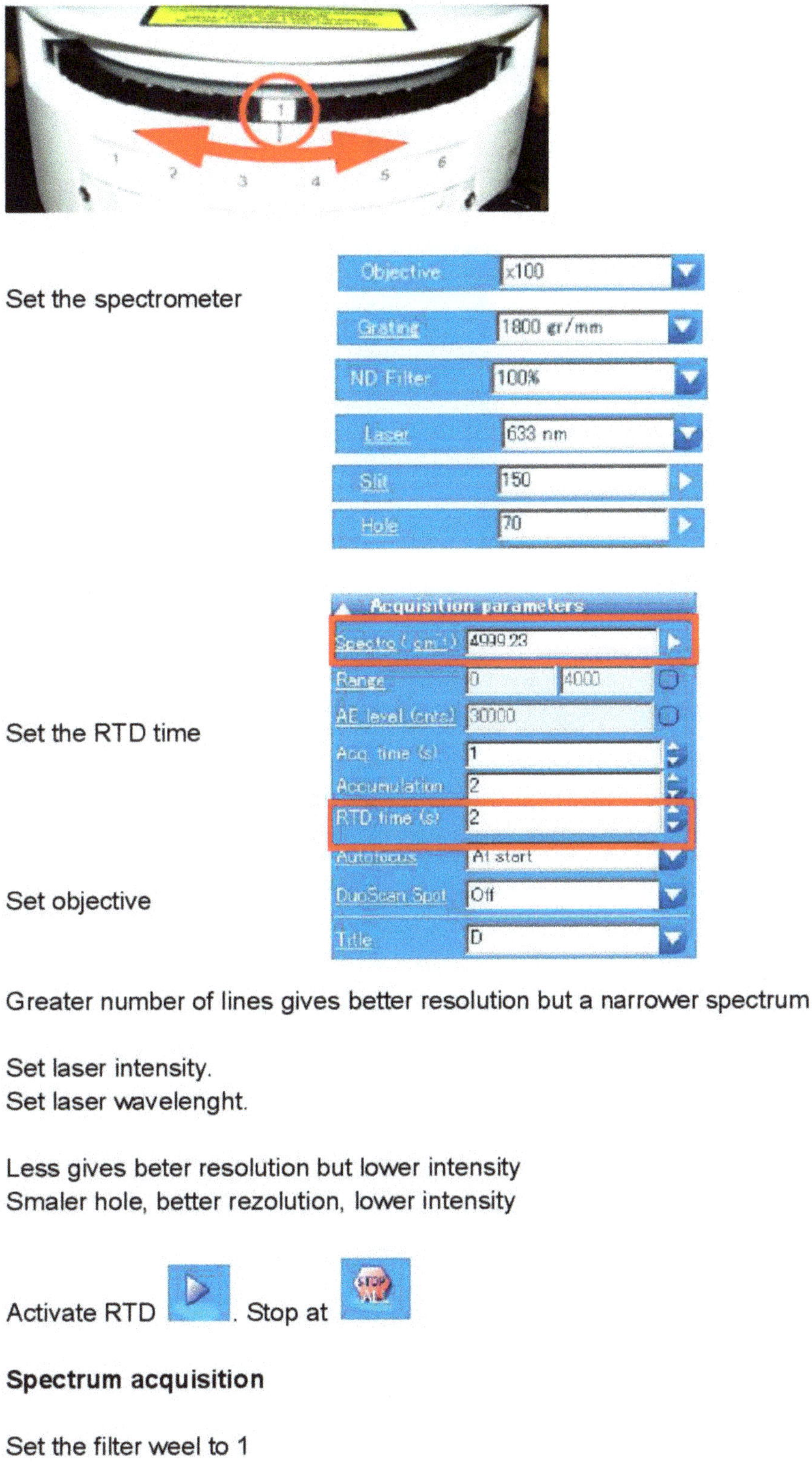

Figure 4: Acquisition parameters.

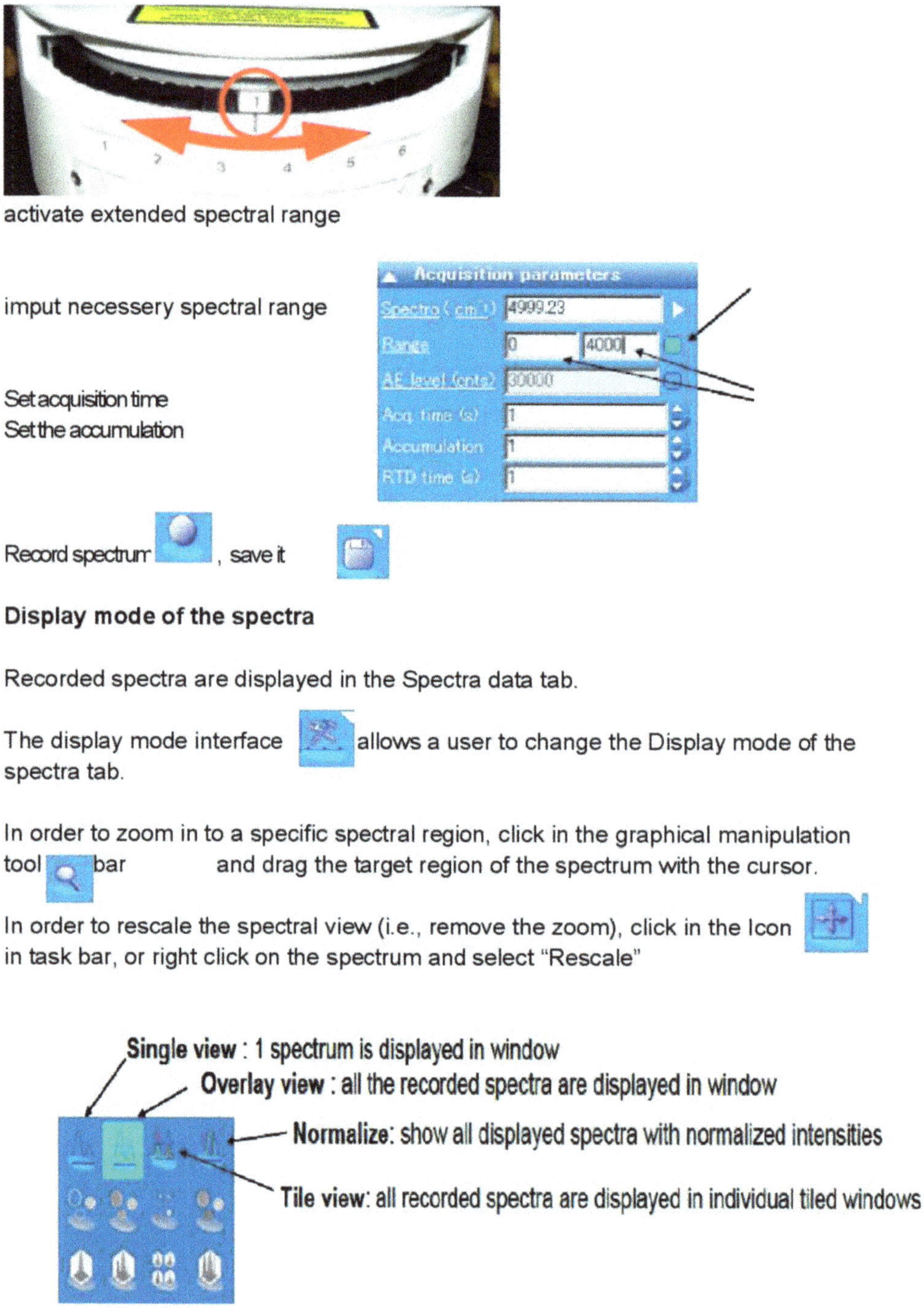

Figure 5: Options for viewing the spectrum.

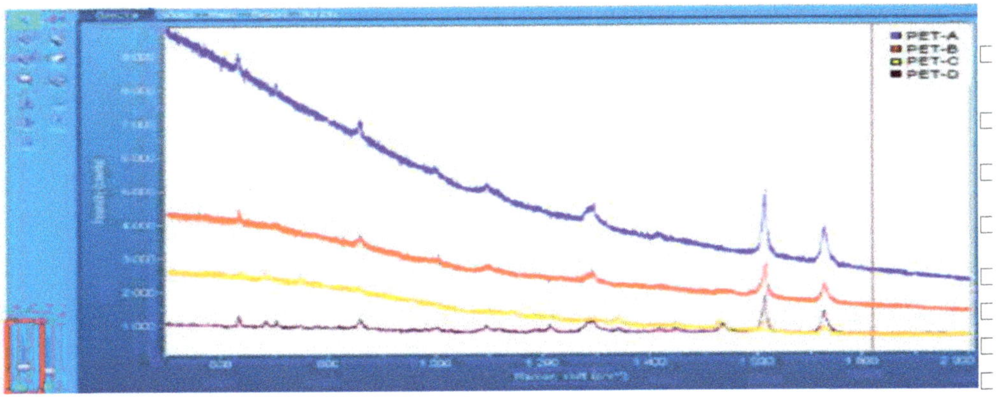

Figure 6: Overlay view of Raman spectra.

www.ingramcontent.com/pod-product-compliance
Ingram Content Group UK Ltd.
Pitfield, Milton Keynes, MK11 3LW, UK
UKHW052007051026
15624UKWH00003B/609